# The Air around Us

T. J. Chandler

# The Air
# around Us

Aldus Books London

| | |
|---|---|
| Editors | Joanna Evans |
| | Ken Coton |
| Designer | David Cox |
| Assistants | David Nash |
| | Richard Hall |
| Research | Tony Walker |

SBN: 490 00081 9

First published in 1967 by
Aldus Books Limited
Aldus House, Fitzroy Square, London W1
Distributed in the United Kingdom
and the Commonwealth by
W. H. Allen & Company
43 Essex Street, London WC2

Printed in Italy by Arnoldo Mondadori Verona

# Contents

# 1  Observing the Weather

No study of the earth would be complete without some knowledge of the atmosphere that surrounds it. For without the mixture of gases that it contains, no plant or animal life could exist; while its behaviour—which we experience as weather and climate—affects our lives in many important ways.

Most obviously, climate—which can be defined as an average of the types of weather experienced in a region over a long period—has helped to form the natural landscape. Over the centuries, it has shaped the land into hills and valleys; it has established and controlled the flow and volume of rivers, lakes, and seas; it has decided the positions of deserts and ice-caps, and determined the types of vegetation and their distribution.

Over shorter periods, climate and weather have also greatly influenced the way in which men live both as individuals and as members of a community. On them depend not only the basic human needs for food, shelter, and clothing, but also such aspects of day-to-day life as man's demand for fuel, the safety of his transport, and even his physical and mental well-being. Clearly, then, as we learn more about climate and weather, we are in a better position to exploit them to our advantage. In short, a knowledge of *meteorology*—as the study of the atmosphere is called—can contribute enormously to man's constant search for better living.

In the past, man soon learned that some knowledge of the weather was essential for his survival. But no real understanding of the weather could be achieved without scientific study of the atmosphere—a study that

No real scientific study of climate and weather was possible until the development of accurate measuring instruments, beginning in the 17th century. Before then, fanciful theories about the earth and its climate were common—like the 16th-century drawing, above, showing the supposed three regions of air. Right, 17th-century painting by J. Porcellis, entitled *Dutch Ships in a Gale*—a forceful reminder of just one way in which weather influences man's life.

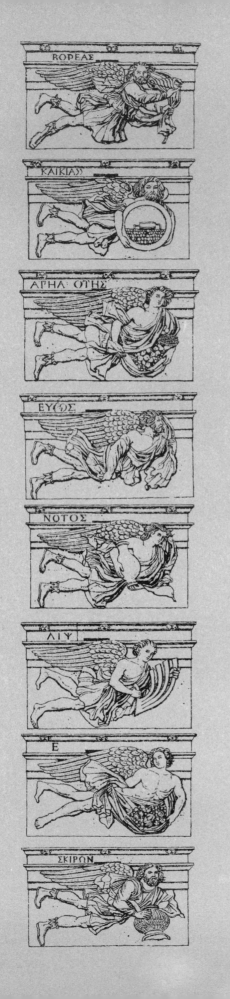

depended on *measuring* the atmospheric components. The development of meteorology therefore goes hand in hand with the history of its instruments, because until these were invented, men could only gaze and wonder at the sky above them.

The earliest weather prophets, who lived in the Middle and Far East some 4000 years ago, related the weather to the positions of stars and planets. Other ancient peoples believed that the weather was governed by the caprice of their various gods. So the Greeks who lived in Homer's time (around 900 B.C.) thought that a thunderbolt was a sign of Zeus's displeasure, while a storm at sea was raised by an angry Poseidon. In the Bible, too, the weather is seen as a manifestation of the divine will, though there are references—in the book of Job, for instance —that show a remarkable knowledge of weather lore. But, as in so many other fields, the first people to develop a scientific approach to the weather were the Greeks of the period known as the Enlightenment. Aristotle (who lived in the fourth century B.C.) was the most notable of these pioneers; his *Meteorologica,* for example, gives fairly accurate descriptions of such processes as the formation of dew, hoar frost, and rainbows. Unfortunately, it was impossible to check or develop his ideas, because no instruments then existed for measuring temperature, pressure, and humidity.

Left, drawings of figures personifying the eight winds of the ancient Greeks carved on the Tower of Winds, Athens (100 B.C.). They are Boreas (north wind); Kaikias (north-east); Apeliotes (east); Eurus (south-east); Notos (south); Lips (south-west); Zephyros (west); and Skiron (north-west).

Opposite page: left, the barometer described by Torricelli in 1643. A glass tube (B) filled with mercury is placed in a bowl of mercury; the mercury in the tube then drops to the level at which it is supported by the atmospheric pressure of the mercury in the dish. (Tube A repeats the experiment in a different form.) Centre, a mercury barometer made from Torricelli's specifications. Right, an 18th-century hygrometer. As the hair expands with moisture, the pointer registers the degree of atmospheric humidity.

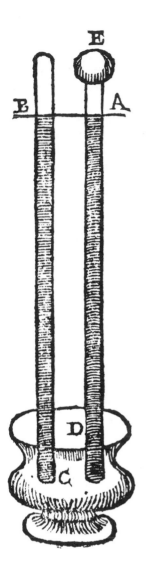

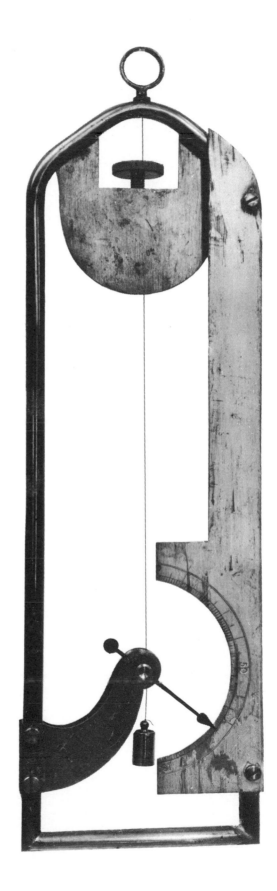

13

WEATHER REPORT.

Sept. 3rd        1860.

At 9 A.M.

| | B. | B. | M. | D. | F. | C. | I. |
|---|---|---|---|---|---|---|---|
| Aberdeen | | | | | | | |
| Greenock | 30·07 | 55 | 52 | WSW | 2 | 1 | b |
| Berwick | | | | | | | |
| Copenhagen | | | | | | | |
| Portrush | | | | | | | |
| Hull | 30·06 | 54 | 52 | W | 2 | 1 | o |
| Liverpool | | | | | | | |
| Queenstown | | | | | | | |
| Helder | | | | | | | |
| Yarmouth | 30·06 | 63 | 59 | NW | 2 | 5 | c |
| London | 30·13 | 58 | 54 | W | 2 | 2 | b |
| Dunkirk | 30·15 | 59 | 52 | WSW | 0 | 1 | b |
| Dover | | | | | | | |
| Portsmouth | 29·96 | 59 | 58 | SW | 3 | 3 | bc |
| Plymouth | 30·06 | 60 | — | NNW | 2 | 0 | oc |
| Cherbourg | 30·11 | 61 | 55 | WNW | — | 1 | bc |
| Penzance Havre | | 57 | — | — | — | 2 | bc |
| Jersey | 30·15 | 59 | 56 | NNW | 2 | 3 | bc |
| Brest | 30·07 | 52 | — | NW | 0 | 9 | oc |
| Bayonne | | | | | | | |
| Lisbon | | | | | | | |

EXPLANATION.

B.—Barometer corrected and reduced to 32° at sea-level (mean).  B.—Exposed (but shaded) thermometer.

M.—Moistened bulb (for evaporation and dew point).  D.—Direction of wind (true).  F.—Force (0 to 12).

C.—Cloud (1 to 9) proportion.  I.—Initial letters: b.—blue sky; c.—clouds (detached); f.—fog; h.—hail;

l.—lightning; m.—misty (hazy); o.—overcast (dull); r.—rain; s.—snow; t.—thunder.

NOTE.—A letter repeated augments—thus, r r much rain.

Above, the first official weather report issued by the Meteorological Department of the Board of Trade. Such reports, mainly intended for shipping, were displayed at ports and at Lloyd's.

Left, a diagonal barometer made in London in 1753. The upper part of the tube is bent at an angle, so that the mercury in it moves a greater distance for a given pressure change, thus providing a more clearly readable scale. Below the barometer is a hygroscope, which registers variations in humidity, and on the right is a Fahrenheit thermometer. The central panel consists of a "perpetual" calendar for a hundred years.

Lack of instruments meant that there was almost no development in the scientific understanding of the weather from the time of Aristotle to the 17th century. In the meantime, there grew up a vast unrelated jumble of astrological theories and local traditions about the weather. Nevertheless, this traditional lore, usually expressed in the form of rhymes or jingles, often reflected a sound understanding of local weather, which is hardly surprising among mainly country people, who were more accustomed than modern city dwellers to observe the weather and far more dependent on it for their livelihood. Even so, the accuracy of these traditional sayings was sometimes sacrificed to the demands of scansion and rhyme.

Meteorology as a science was born in 1643, with the invention of the simple barometer by the Italian, Evangelista Torricelli. Until then, no one had really been aware of the existence of atmospheric pressure, still less of its association with the weather. It was not long, however, before it was seen that fluctuations in pressure were related to weather changes.

In 1670, the first "weather glass" was made by the English scientist Robert Hooke, with low pressure equated with rain and stormy conditions, and high pressure with dry, fair conditions. (These terms still appear on the face of "hall barometers," showing a continuing but misplaced confidence in this simple association.) And about the same time as the barometer, the first liquid-in-glass thermometer was invented (probably by Galileo), and also the first hygrometer, which used a human hair (which expands with moisture) to measure atmospheric humidity.

Three basic elements of the atmosphere—pressure, temperature, and humidity—could now be measured, and scientists began for the first time to work out the relationships between these and other atmospheric features such as winds, rainfall, sunshine, and clouds. But the observations they made were still scattered and unco-ordinated, and so failed to produce any overall picture of the weather covering large areas of land and sea. The first weather map of which we have any record is that made by the English astronomer Edmond Halley in 1686. Halley charted the winds between about 30° N and S, but only gave the average conditions over a long period of time. Weather charts showing simultaneous conditions over a large area had to wait almost two centuries for the rapid transfer of information made possible by the invention of the electric telegraph in the 1840s. (Our present weather maps are the successors of those sold for a penny at the Great Exhibition in London in 1851.) This major advance in methods of communication marked the beginning of *synoptic* meteorology—the comparative study of weather conditions over a large area—and many countries soon set up organizations for weather forecasting and research. In Britain, for example, Admiral Robert Fitzroy was in 1854 appointed head of the new Meteorological Department of the Board of Trade, whose particular task was the issue of gale warnings to merchant ships.

At the same time, more stations were set up for recording weather. Simple measurements of air temperature began to be kept in the 17th century, but it was not until the 19th

century that continuous records were made for the more accurate charting of weather. During the present century the number of weather stations has increased enormously, but even so there are still large areas of the world where they are few and far between. Few exist, for example, in the oceans, although several countries have co-operated in the last 20 years to establish a network of weather ships, particularly outside the main shipping routes. On the routes themselves, a great deal of information is received from merchant and passenger ships. Nevertheless, our knowledge of the weather in the world's sea areas (which cover almost three quarters of the earth's surface) is still scanty.

All recording stations aim to provide information for weather forecasting and at the same time to build up a record of climate. At stations that provide detailed synoptic information, readings are normally taken hourly; at others, the number of daily observations varies with the station's function. But however often observations are made,

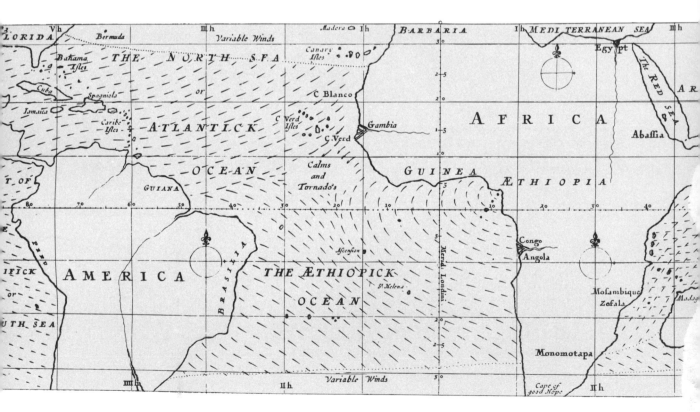

they usually follow a standard procedure. First, a few minutes before the hour, an observer notes the state of the sky; the amount, type, and height of the clouds; whether rain, snow, or hail is falling; the visibility; and the state of the ground. Once a day he measures the amount of sunshine in the past 24 hours, and estimates (or measures if he has instruments) the direction and strength of the wind. He then opens the thermometer screen, usually a louvred wooden box painted white, to take the temperature readings. Most screens contain four thermometers. One of these is the "dry-bulb" thermometer, which gives the ordinary air temperature. Another is the "wet-bulb" thermometer whose bulb is encased in wet muslin. Because water on the muslin evaporates, the wet-bulb thermometer shows a lower temperature than the dry-bulb. The difference between the two readings provides a measure of the relative humidity of the air. The other two thermometers register the maximum and minimum temperatures that have occurred

Above right, a standard thermometer screen at Kew. At four feet from the ground are four thermometers, a thermograph (which keeps a continuous record of temperature), and a hygrograph (which registers humidity changes). Right, close-up of a thermometer screen. One of the two vertical thermometers— the dry bulb—shows the ordinary air temperature; the other—the wet bulb—has its bulb covered with wet muslin. The difference in temperature of the two thermometers provides data from which the air's relative humidity is calculated. The two horizontal thermometers indicate the maximum and minimum temperatures since the last reading.

Left, part of the map compiled by the English astronomer Edmund Halley in 1686, showing the winds in the tropics. The map illustrated an article in which Halley described the trade winds and doldrums, and correctly attributed them to the ascent of warm air at the equator.

Delegates to the International Meteorological Congress, Rome, in 1879. Several conferences were held at this time to establish the procedure for exchanging weather information.

since the last observation. Finally, the observer measures the amount of rain or snow collected by the rain gauge, and notes the degree of atmospheric pressure shown by the barometer and the pressure changes recorded by the barograph.

Once assembled, all the information is converted into an internationally recognized numerical code and sent by telephone, telegraph, or wireless to the central meteorological offices in different countries. Transmission is swiftly followed by an exchange of readings between the various central offices, so that within an hour of observation a picture has been built up of the weather over a very large proportion of the hemisphere. This service for exchanging weather information is run by an international body, the World Meteorological Organization, whose headquarters are in Geneva. More recently, the inter-office exchange of information has been streamlined even further by the long-distance transmission between central offices of facsimiles of weather maps already marked with station information.

The meteorologist's greatest problem has always been to find ways of observing and studying atmospheric conditions above the earth's surface. Early meteorologists tried to overcome this difficulty by making use of mountain observatories. These were, however, rare and inaccessible, and in any case do not truly represent conditions in the free atmosphere. So, in the late 19th century, meteorologists began to send up manned balloons to observe the first few thousand feet of the earth's atmosphere. These were soon replaced by unmanned balloons able to carry lightweight recording instruments high above the level at which a man can survive unprotected. Unfortunately, since there was often a long wait—sometimes of months or even years—between the release of a balloon and the return of its recorders, this technique could not be used for daily analyses.

Some of these early problems were solved by the aeroplane, and during the 1920s many countries began to organize regular meteorological flights to measure temperature, pressure, and humidity, and to observe such

Above, French scientists observing the atmosphere from a balloon in 1875. Such balloon ascents were the first attempts to study the atmosphere in depth.

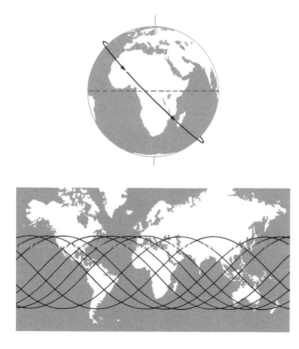

Today our knowledge of the upper atmosphere has been enormously extended by information received from weather satellites. Top, the orbit of a weather satellite travelling around the earth at an angle of 45° to the equator. Because of the earth's rotation, the satellite would describe the wavy paths shown on the map, above. Only the arctic and antarctic regions are therefore beyond its range of view.

things as clouds and winds up to several thousand feet above the earth. Since the late 1930s, however, a simpler and cheaper method of aerial weather observation has been developed: the use of sounding balloons that, rising as high as 100,000 feet, transmit instantaneous readings to radio receivers on the ground. Modern meteorological rockets climb even higher, though their performance pales beside the spectacular advance in weather mapping, analysis, and forecasting brought about by weather satellites.

Satellites view the atmosphere from a height of about 450 miles and with their sensors register both direct and scattered radiation from the sun above and from the earth and atmosphere below. This radiation ranges in length from the shortest, invisible ultra-violet wave-lengths, through the visible part of the spectrum, to the longest, again invisible, infra-red bands. And so, with instruments sensitive to different parts of the spectrum, it is possible to identify and measure such substances as ozone, water vapour, and carbon dioxide by the amount and type of radiation they emit or absorb. Other instruments record cloud patterns. Satellites register also such non-meteorological data as the distribution on land of snow and ice, crops and natural vegetation, and even the location of various soil types and minerals.

All this information is sent automatically to ground receivers, either immediately or after a period of storage. The contribution of satellite observation to meteorology, and indeed to all the earth sciences, is invaluable, particularly where the satellites are in polar orbit—that is, orbiting from north to south so that they cover the whole of the earth rotating beneath them.

Today it is possible to explore the atmosphere in far greater depth than ever before. And as in so many other sciences, some preconceived meteorological ideas have had to be revised in the light of new knowledge. For the rest of this book, we shall try to outline some of the basic findings of meteorology: why the atmosphere behaves as it does and how this behaviour produces the familiar features of sun, rain, and wind that make up the patterns of our weather and climate.

# 2 Structure of the Atmosphere

The ocean of air in which we live reaches upward to a height that has been somewhat arbitrarily fixed at 600 miles (1000 kilometres). And in many ways, this all-pervading mass of air does behave like the waters of the sea. In some places it moves extremely slowly, while in others it is violently agitated. Again, like the creatures that crawl on the ocean bed, animals that live at the bottom of the atmosphere are subject to considerable pressure. At ground level the average pressure of air is 14.7 lbs per square inch, so its total pressure on the human body is about 10 tons. For in the atmosphere, as in the sea, the pressure at the bottom is much greater than it is near the top. The effect of gravity is to concentrate the atmosphere nearest the earth, so there is a rapid fall in air density and pressure in the first few thousand feet; at greater heights, the decrease gradually slackens. Meteorologists measure these changes in units of atmospheric pressure called *millibars*. One millibar equals one thousand dynes per square centimetre. (A dyne is the force required to move one gram of mass one centimetre per second.) The normal pressure of the atmosphere is 1013 millibars.

As a result of this compression, about one half of the atmosphere's mass lies in the $3\frac{1}{2}$ miles nearest the earth, and more than 99 per cent within 25 miles. At 60 miles, the air is so rarefied that it is almost a vacuum, and has only one millionth of the pressure that it had at ground level. Higher still, manned satellites, which circle the earth at heights of 150 to 300 miles, and weather satellites, at about 450 miles, are almost entirely free from frictional drag and heating.

The physical behaviour of the atmosphere is extremely complicated, and still something of a mystery. One reason for our relative ignorance about the atmosphere is its size: in the ordinary way, chemists and physicists deal with material that they can handle in laboratory experiments, but the meteorologist must study the whole atmosphere—a vast area whose individual features he is unable to isolate or control. Another difficulty is the irregularity of the atmosphere's behaviour: no two days, months, years, or centuries have the same cycles of weather. Yet the gases, liquids, and solids in the atmosphere are governed by chemical and physical laws, as they are anywhere else. It is the number of these

A multiple-exposure photograph shows how an American *Nimbus* satellite unfolds its "wings" in orbit. The satellite, which is powered by electricity from solar radiation, transmits a continuous record of world-wide weather conditions from a height of about 450 miles.

laws and their complicated interaction that makes the behaviour of the atmosphere difficult to understand and even more difficult to predict. Our limited, "bottom-of-the-tank" view of the air above us has, however, been enormously extended during the last 60 years by information received from balloons, aeroplanes, rockets, and satellites.

Yet though the atmosphere behaves in a complex manner, its chemical make-up is comparatively simple. It is a mixture of two types of gases: first, the permanent gases, of which nitrogen and oxygen make up 99 per cent and trace gases 1 per cent; second, the variable gases, some of which occur naturally and others as the result of local conditions such as fuel combustion. The most important of the variable gases are ozone and water vapour. Up to about 40 miles, the relative quantities of the permanent gases are more or less constant. The distribution of ozone and water vapour, however, is very uneven.

Ozone is mostly found at a very high level, where it is created by a series of chemical reactions that take place only by the agency of light, and are known as *photochemical* reactions. First, the strong ultra-violet radiation from the sun dissociates some of the molecular oxygen into atomic oxygen. This atomic oxygen (O) combines with existing molecular oxygen ($O_2$) to form ozone ($O_3$). At the same time, the opposite process is taking place, also by photochemical means, as ozone is dissociated by solar radiation of longer wavelength (still in the ultra-violet range). The quantity of ozone therefore depends on the relative intensity of three processes: first, the break-down of molecular oxygen into atomic oxygen; second, the combination of molecular and atomic oxygen to form ozone; and third, the destruction of ozone. At 50 miles and more above the earth, ozone is more easily destroyed than produced, and so there is no ozone; at heights between 20 and 50 miles, ozone both forms and breaks down again very rapidly; concentrations reach a maximum at about 22 miles. Below 6 miles, molecular oxygen cannot be broken down at all, since all the short-wave solar radiation necessary for its dissociation has already been absorbed at higher levels. So there is no

manufacture of ozone at these low levels, though some is brought down by descending currents of air. These airborne concentrations are generally small, though they may vary from place to place. For example, ozone is an active oxidizing agent, which attacks organic matter. So, since there is less organic matter to consume ozone over water than on land, sea breezes contain slightly larger amounts of this gas—which, despite its popular reputation, is in fact more of a poison than a tonic. Admittedly ozone is a powerful germicide, but any concentration large enough to kill bacteria is intensely irritating to the membranes of the nose and throat.

Below, tables of the two classes of gases in the atmosphere: first, the relative percentages of the permanent gases; and second, the other gases, which are distributed very unevenly.

## PERMANENT GASES %

| nitrogen | 78.084 |
|---|---|
| oxygen | 20.964 |
| argon | 0.934 |
| krypton | < 0.001 |
| hydrogen | < 0.001 |
| xenon | < 0.001 |
| neon | < 0.001 |
| helium | < 0.001 |
| methane | < 0.001 |
| nitrous oxide | < 0.001 |

## OTHER GASES variable

water vapour
carbon dioxide
ozone
hydrogen peroxide
ammonia
hydrogen sulphide
sulphur dioxide
sulphur trioxide
carbon monoxide
radon

Unlike ozone, water is familiar to us—as a liquid, including the small droplets of steam and most clouds, or frozen as ice and snow—so we sometimes forget its presence as an invisible gas everywhere in the air around us. In fact, though concentrations of water vapour vary enormously in different areas of the world, the average amount of moisture in the air, visible and invisible, is equal to one inch of rain, or 900 tons per acre. Water vapour is nearly all found in the six to nine miles nearest the earth. Higher up, the air is very dry, with few clouds except the rare "mother-of-pearl" clouds at heights of about 16 miles (where the amount of vapour seems to increase a little) and the clouds that shine after dark on a clear night—known as *noctilucent* clouds—at heights of about 50 miles. Rockets fired from Sweden in 1963 showed these very high clouds to be made up of ice-coated dust particles from outer space.

The origin of the water vapour in the atmosphere is uncertain, but one suggestion is that it was created by photochemical processes (probably on methane) similar to those that produce ozone. This would mean that ultra-violet rays caused methane ($CH_4$) to break down into separate hydrogen and carbon atoms. The hydrogen then combined with atomic or molecular oxygen to form $H_2O$.

Below, a diagram representing the manufacture of ozone in the atmosphere. Ultra-violet solar radiation (the blue arrows) dissociates some molecular oxygen ($O_2$) into separate atoms (O). This atomic oxygen then combines with remaining oxygen molecules to form ozone ($O_3$).

Bottom, a Belgian beach scene in 1914. Visitors to the seaside who think they are breathing in health-giving ozone are in fact probably smelling iodine released by rotting seaweed. Generally, little ozone reaches the troposphere, though there is more in the air over the sea than over the land.

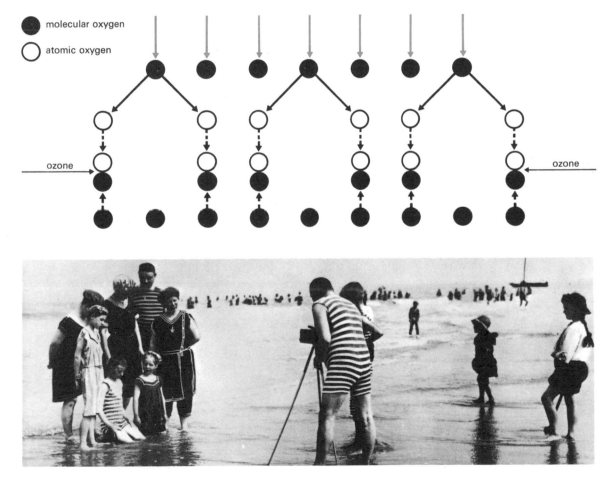

molecular oxygen

atomic oxygen

ozone

ozone

The combination of gases that makes up the atmosphere is largely kept in place by the gravitational pull of the earth. Even so, a continuous slight erosion of the fringes of the atmosphere probably does occur. This happens when gas molecules, which are constantly and rapidly moving in random directions, travel outward at a velocity that overrides the gravitational pull inward of the earth: this velocity is known as the *escape velocity*. The velocity of molecules in a gas increases with temperature; so, by relating the molecular weight of a gas and its temperature, it is possible to work out its chances of escaping from the earth's atmosphere. It is usually estimated that the temperature at 185 miles is approximately 1315°C. Calculations based on this figure show that it would take only about 4000 years for most of the hydrogen molecules to escape. On the other hand, it would take many times longer than the assumed age of the solar system (5000 million years) for the much heavier gases, such as nitrogen and oxygen, to disappear.

Gases that surround a body of less gravitational attraction than the earth obviously escape at a lower velocity. It is therefore not surprising that we cannot detect any atmosphere on the moon, whose force of gravity at its surface is only one sixth that of the earth. Similarly, Mercury—a small, hot planet—has almost no atmosphere, while Jupiter—a huge, cold planet—has a deep, dense atmosphere.

Even so, the mechanism of escape velocity, though important, does not entirely explain the difference in chemical composition of the various planetary atmospheres. For example, the proportion of rare gases to the common gases—nitrogen, oxygen, carbon dioxide, and water vapour—is far less in the terrestrial atmosphere than in the atmosphere of any other planet. Why, then, is the earth's

A photograph taken in August 1965 by an American astronaut from *Gemini V* as it passed over New Guinea at a height of 100 miles. It clearly shows the sharp break between the clouds in the shallow, water-vapour-laden layer immediately surrounding the earth (the troposphere) and the dry, cloudless conditions of the air above.

24

mantle of gases so different? To answer this question, we must try to reconstruct what happened when the earth first came into existence, some 5000 million years ago.

If, as most scientists agree, the newly formed earth was extremely hot—about 9000°C—most of the gases that made up the original atmosphere must have escaped. But as the earth's surface gradually cooled and solidified, other gases that had been dissolved in the earth's liquid crust would have emerged. These gases would have been similar to those now emitted by volcanoes—that is, carbon dioxide, nitrogen, and water vapour. (Today this water vapour may also be a product of existing subterranean water that has percolated into the volcanoes.) As the cooling process continued and the temperature fell below 100°C, much of the water vapour would have condensed and fallen on

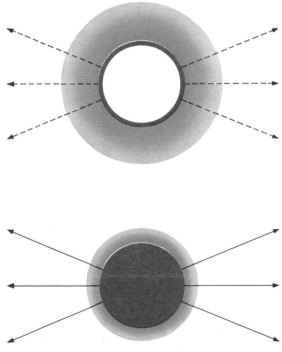

A gas molecule escapes from the atmosphere only if its velocity is great enough to override the gravitational attraction of the earth. Since velocity increases with temperature, a particular gas may stay anchored in the atmosphere of a cool planet (upper), whereas it would escape from a hotter planet (lower) of the same mass.

the earth to form rivers, lakes, and oceans. At roughly the same time, a large proportion of the carbon dioxide in the air would have disappeared by dissolving in the oceans, where some would have attacked calcium salts to form limestone and dolomite rock, which are made up of carbonates of magnesium and calcium. There is also an enormous amount of carbon dissolved in the sea in the form of carbon dioxide, which is the basis of all vegetable—and so animal—life.

The main structure of the earth's atmosphere would have been completed at this stage, with one very important exception: there was still no oxygen. The origin of this gas, the second most abundant in the atmosphere and the one most vital to living things, is still something of a puzzle. One theory is that oxygen was first formed by the splitting of water vapour into atomic hydrogen and atomic oxygen by the photochemical action of the sun. The lighter and more mobile hydrogen then escaped into outer space, while the heavier oxygen atoms were retained by the earth's gravity. Calculations show, however, that this process cannot account for more than a very small amount of the atmosphere's oxygen.

Another explanation is that oxygen was produced by photosynthesis in primitive cell life at a later stage in the earth's development. This probably took place in shallow pools until sufficient oxygen and ozone were built up for plants to spread over the land. When this happened, there would have been a rapid increase in atmospheric oxygen.

Opinion is divided on the nature and extent of changes in the composition of the atmosphere during the last 1000 million years. Geological processes, such as volcanic action and the deposition of limestone and coal, must have had some effect, and there is reason for thinking that during the past 300 million years, oxygen and carbon dioxide have fluctuated above and below their present levels. Even during the last century, there have been small but significant alterations to the atmosphere's delicate chemical balance. For example, the amount of carbon dioxide seems to have risen by as much as 12 per cent since 1900. This rise is probably due largely to the increasing consumption of carbon fuels —mainly coal and oil—which has doubled every 10 years since 1900. Since coal is 90 per cent carbon, its combustion with oxygen leads to an increase of carbon dioxide. When oil, which is a hydrocarbon (that is, containing both carbon and hydrogen) is burned, the resulting products are carbon dioxide and water. Increased combustion has also produced concentrations of such noxious gases as carbon monoxide from internal-combustion engines and sulphur dioxide from oil-powered furnaces. Even more ominous is the increasing amount of radio-active fall-out from atomic explosions. (Many radio-active substances, however, have only a short life in the lower atmosphere; they often rise to quite high levels, where they may circulate for many years before decaying or returning to earth.) And as the space race accelerates, there is a growing danger that exhaust gases from rocket motors may disturb the very delicate chemical balance of the ozone layer.

Such recent minor changes may seem negligible when set against the stability of the atmosphere as a whole. But here it is worth remembering that the rarer gases in the atmosphere can have a far greater influence on climate and weather than their small concentrations might suggest. For example, it is generally agreed that the recent increase in carbon dioxide has helped to alter the balance of heat in the atmosphere by absorbing the infra-red rays of terrestrial radiation. This has in turn reduced the loss of heat from the earth to space, and raised the near-surface air temperatures. Though the precise increase in temperature is still uncertain, it has been estimated at $0.011°C$ a year—an estimate that shows the close and delicate relationship between temperature and the atmosphere's chemical make-up.

Before the end of the 19th century, it was generally thought that air temperatures fell continuously from the bottom to the top of the atmosphere at the rate that had been observed in the lowest layers—that is, at about $1°C$ per 500 feet—eventually reaching absolute zero ($-273°C$) in the interstellar void. This belief was exploded by the French meteorologist Léon Teisserenc de Bort, who

Above, photo of a radiosonde balloon with parachute and equipment beneath. (Distances between components have been greatly reduced.) Below, diagram of a radiosonde, which combines instruments for recording changes of temperature, humidity, and pressure with a radio transmitter.

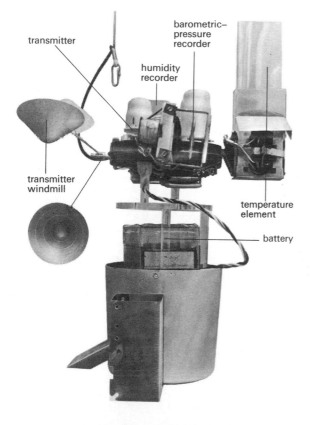

transmitter

barometric–pressure recorder

humidity recorder

transmitter windmill

temperature element

battery

during the 1890s succeeded in sending up balloons with recording instruments to heights of about nine miles. From the information recorded by these instruments, de Bort showed that at about seven miles above Europe there is a sharp break, or discontinuity, above which temperatures stop falling and even increase a little with height. The discontinuity was later found to be a universal feature of the atmosphere, but it varies in height from about five miles over the poles to about ten miles over the equator. In the last few years, daily soundings taken at weather stations (particularly those in the Northern Hemisphere) have shown that the discontinuity—known as the *tropopause*—is not continuous, as was originally supposed: there are generally two, and sometimes three, separate levels of discontinuity in different latitudes in both hemispheres, with substantial breaks and changes of altitude between them. A distinction is therefore usually made between a *tropical* tropopause, generally at 50,000 to 60,000 feet, and a *polar* tropopause at about 30,000 feet. The break between the two boundary surfaces varies from day to day as well as from month to month, but it generally lies between latitudes 35 and 50° N and s. A third tropopause that can sometimes be distinguished from the other two in intermediate latitudes is known as the *mid-latitude* tropopause. This breaks with the tropical tropopause at about 30° and with the polar tropopause at around 45°.

The tropopause separates the lower atmosphere—the *troposphere*—from the next highest layer above it—the *stratosphere*. It was once thought that the heat balance of the troposphere was maintained by convection currents, and that that of the stratosphere was the result of a radiation balance with the sun, space, and earth. The subsequent discovery of very strong winds and rapid changes of temperature in the stratosphere has rather discredited this view, and nobody is now certain why there should be such a sharp discontinuity between the troposphere and the stratosphere. But, whatever the explanation, the differences in the thermal make-up of the two layers are largely responsible for their distinctive meteorology.

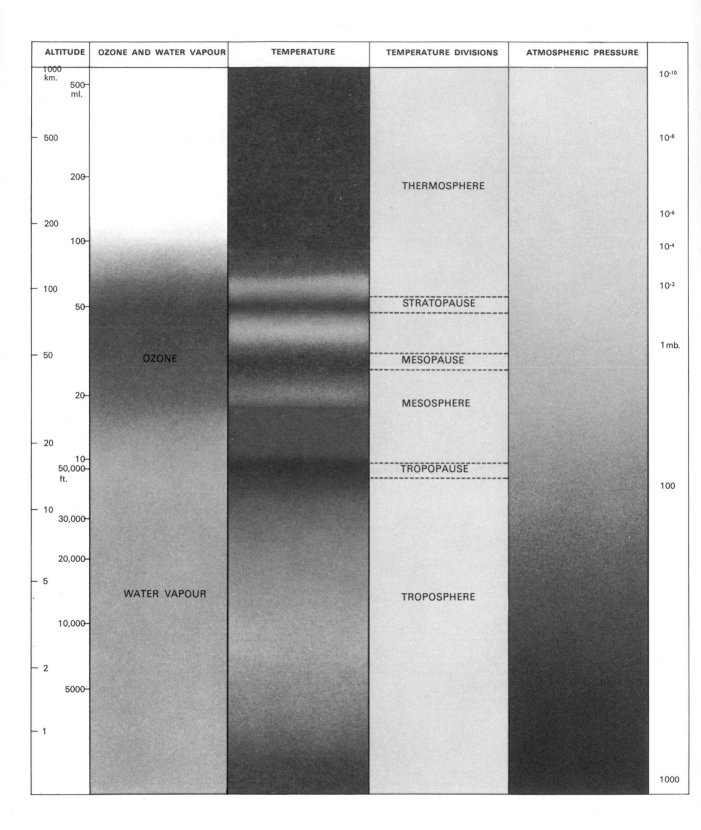

| ALTITUDE | OZONE AND WATER VAPOUR | TEMPERATURE | TEMPERATURE DIVISIONS | ATMOSPHERIC PRESSURE |
|---|---|---|---|---|

1000 km.
500 ml.

500

200

200

100

100

50

OZONE

50

20

20

10
50,000 ft.

10

30,000

20,000

5

WATER VAPOUR

10,000

2

5000

1

THERMOSPHERE

STRATOPAUSE

MESOPAUSE

MESOSPHERE

TROPOPAUSE

TROPOSPHERE

$10^{-10}$

$10^{-8}$

$10^{-6}$

$10^{-4}$

$10^{-2}$

1 mb.

100

1000

Charts of the atmosphere up to a height of 500 miles. Both altitude (left of page) and pressure (right of page) are on a logarithmic scale. First column shows levels at which ozone and water vapour are mainly found.

Second column shows the atmosphere's thermal structure. High-temperature levels are shaded red and low-temperature, blue. Breaks in temperature provide the boundary lines between the layers of the atmosphere, set out in column three.

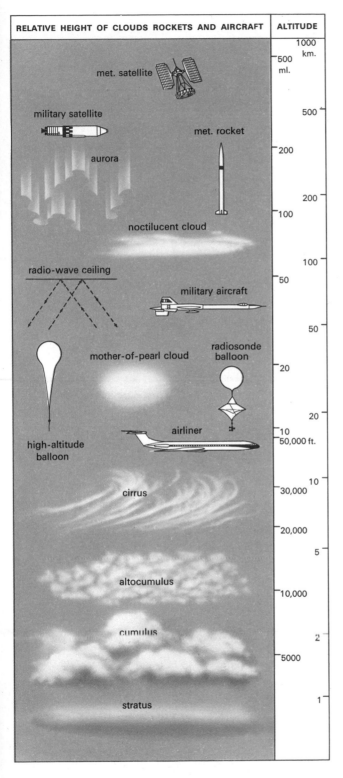

**RELATIVE HEIGHT OF CLOUDS ROCKETS AND AIRCRAFT**

**ALTITUDE**

1000 km.
500 ml.

met. satellite

military satellite

met. rocket

aurora

500

noctilucent cloud

200

100

radio-wave ceiling

100

military aircraft

50

mother-of-pearl cloud

radiosonde balloon

50

high-altitude balloon

airliner

20

20

10
50,000 ft.

cirrus

10
30,000

20,000

altocumulus

5

10,000

cumulus

2

5000

stratus

1

thermosphere

centre of earth

Above right, photographs of some of the clouds and aircraft shown in the above chart. From bottom to top: an airliner flying above the cloud base; a high-flying military aircraft; noctilucent clouds at about 50 miles; and aurora at about 150 miles.

Far right, a diagram relating the 500-mile depth of the atmosphere illustrated on this spread to the earth's 4000-mile radius in order to show the shallowness of the atmosphere compared with the radius of the earth.

Since the troposphere contains nearly all the atmosphere's water vapour, most of our weather is produced by conditions in this comparatively shallow surface layer. And because of the meteorological importance of the troposphere (and the lack of information about the stratosphere), meteorologists largely ignored the higher parts of the atmosphere before World War II. Since 1950, however, our knowledge of the atmosphere above seven miles has increased enormously, largely because of the introduction of high-flying aircraft. Most of the big civil jet airliners fly "above the weather" at heights of 35,000 to 40,000 feet—that is, in the lower stratosphere, above nearly all the clouds and all the precipitation. The new supersonic jets now being built to cross the Atlantic in two to three hours will fly even higher, at 50,000 to 70,000 feet. At this height, which is well into the stratosphere, the air is very thin and produces only a small amount of frictional drag and skin heating. Some military aircraft climb as high as 50 miles (265,000 feet), a height well above the stratosphere.

Most of our information about the first 100,000 feet of the atmosphere is received from radiosondes—radio sensors sent up by balloon to transmit pressure, temperature, and humidity readings. Above this height, meteorological rockets have carried instruments up to 360,000 feet. These rockets release radar reflectors whose descent is then traced by ground radar to make a record of wind speed and direction.

One result of the high-altitude research carried out during the last 20 years is that we can now recognize and define new layers in the upper atmosphere. These layers, all of which have certain distinctive physical and chemical characteristics, are separated from one another by narrow zones of transition, which are known as boundary surfaces or pauses. We have already seen how the troposphere, the layer nearest the earth, is separated from the stratosphere above it by the tropopause at heights varying from 5 to 10 miles. The stratosphere is extremely dry, and has a nearly uniform average temperature of about –60°C in its lower levels; this increases to a maximum of 7 to 18°C in its upper parts,

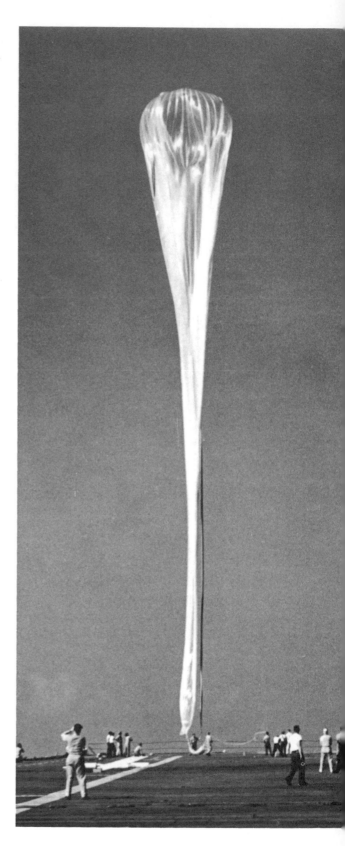

The launching of a balloon carrying radiosonde equipment from an American aircraft carrier in 1960. Such balloons rise about 100,000 feet before bursting, and provide most of our information about the atmosphere up to this height.

at 32 to 37 miles. This consistent thermal maximum is taken as the boundary surface—known as the *stratopause*—that separates the stratosphere from the *mesosphere* above.

The comparatively high temperatures of the stratopause are roughly the same as those at ground level; they probably result from the thorough absorption of the sun's ultra-violet and infra-red radiation by the ozone layer, and it has recently been found that sudden warmings occur at these levels. Within the mesosphere, however, the amount of ozone falls off sharply. Mean temperatures are therefore far lower—about –70°C at about 50 miles—than they are in the stratosphere below. This consistent temperature minimum, known as the *mesopause,* separates the mesosphere from the highest layer yet recognized, the *thermosphere.*

An alternative name for the thermosphere is the *ionosphere,* because at this height the atmospheric gases become ionized—that is, dissociated into individual, electrically charged particles. Since these ionized particles reflect electro-magnetic waves, the thermosphere is of great practical value because it makes possible all long-distance radio communication without satellite. Yet, apart from its efficiency as a mirror for radio waves, we know very little about the thermosphere. It seems that temperature once more increases with height as a result of absorption by molecular oxygen and nitrogen of ultraviolet radiation. Recent temperature measurements made by satellite at heights of about 63 miles—well into the thermosphere—have been very high. Above the equator they were about 925°C, and as much as 1480°C above the North Pole. Still, as more information based on high-rocket soundings and weather-satellite readings builds up, we shall probably not only learn much more about the thermosphere but also be able to plot new layers in the uncharted regions beyond it.

Below, the Skua meteorological rocket, which reaches a height of over 60 miles. This British rocket is one of many recently developed for recording conditions in the upper atmosphere.

Diagram shows rocket's general design; it includes both radiosonde equipment and a radar-reflective parachute for tracking wind. Bottom, the parachute and sonde assembled in the nose cone.

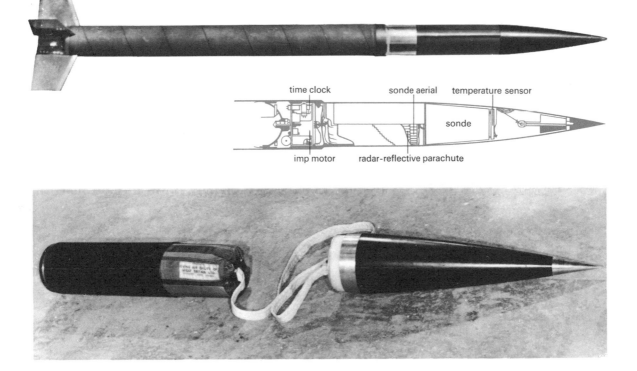

time clock    sonde aerial    temperature sensor

sonde

imp motor    radar-reflective parachute

# 3 Heat Balance of Earth and Atmosphere

Of all the planets in the solar system, the earth is the only one with a moderate temperature, neither very hot nor extremely cold; the average temperature of its surface is approximately 10°C. Since this figure has remained fairly constant over the centuries in spite of small annual changes it follows that the earth and its atmosphere must return to space roughly the same amount of energy as they receive from the sun. Without this balance, the earth's temperatures, and therefore climates, would progressively change. (As we shall see in Chapter 12, such fluctuations have occurred in the past, but at present there is clearly a roughly equal input and output of energy if temperatures are averaged out over a period of years.) The achievement of this balance of energy involves several very important heat exchanges, all of which are crucial to the study of meteorology.

The sun is the original source of almost all the earth's energy. Any other star or planet or the interior of the earth itself provides so little heat in comparison that they can be discounted in our calculations. Although among the billions of stars in our own galaxy the sun is only average in size, it still has a diameter 110 times greater than that of the earth. Because of this—and because of its relative closeness, which varies from 91 to 94 million miles—it is the earth's dominant source of energy. Even so, only about 1 two-thousand-millionth of the energy emitted by the sun reaches the fringes of the earth's atmosphere; the rest is lost or absorbed elsewhere in space.

The sun is a gaseous sphere with a surface temperature of about 6275°C. It is believed that its energy is created by the fusion of hydrogen atoms to form helium—a process that takes place deep in the sun's interior. The energy is then radiated and convected to the sun's surface, from where it is emitted across space, finally reaching the earth in two forms: one is electro-magnetic waves, and the other is a stream of ionized particles. Once it has been absorbed, this energy is reconverted into heat, which warms the earth. The earth, in its turn, also radiates energy, but the wave-length and amount of this energy depend on the type and temperature of its own surface. Any body transmits most of its radiant energy at a wave-length inversely proportional to its surface temperature; in other words, the higher the temperature, the shorter the wave-length of maximum transmission. Furthermore, the *total* amount of energy radiated by a body is proportional to the fourth power of its *absolute* temperature—that is, its actual temperature plus 273°C. So the sun, at very high temperatures, emits enormous quantities

The sun is the earth's dominant source of heat and light. Incoming solar radiation is balanced by the energy returned to space by the earth and atmosphere. This balance of energy is achieved by several very important heat exchanges, which are discussed in this chapter.

## ISOTHERMS OF JANUARY TEMPERATURES

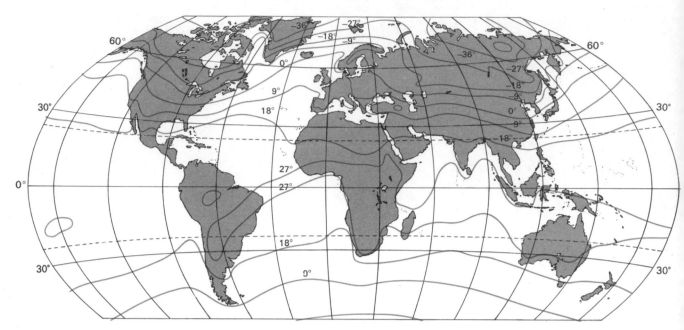

The intensity of solar radiation received at any point on the earth's surface depends partly on the angle of incidence; this varies with the latitude. Right, a diagram of the solar beam at midday at the spring and autumn equinoxes, when it is roughly at right angles to the equator (A). In higher latitudes, it strikes the earth more obliquely, and therefore shines less intensely over a larger area (B and C).

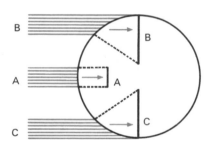

## ISOTHERMS OF JULY TEMPERATURES

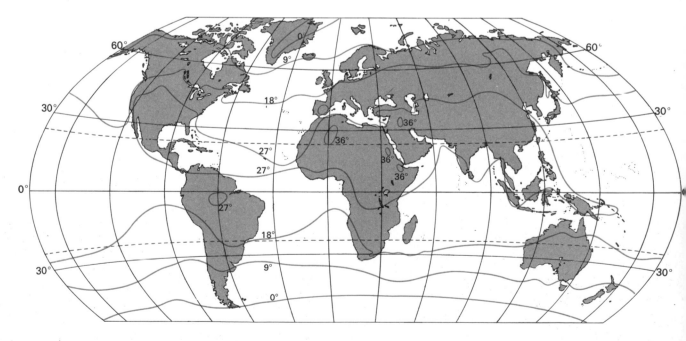

of mainly short-wave radiation, while the much cooler earth radiates smaller amounts of mainly long-wave radiation. Ninety-nine per cent of the sun's radiation has a wave-length of between 0.15 and 4 microns. (A micron is one millionth of a metre.) The maximum transmission of radiation occurs at a wave-length of about 0.47 microns, and about half the sun's total radiation is in the waveband that is visible to the human eye as light—that is, between 0.38 and 0.77 microns. Most of the remaining radiation is made up of the slightly shorter (ultra-violet) and the slightly longer (infra-red) rays.

The amount of solar radiation received at the fringes of the earth's atmosphere depends mostly on the angle made by the sun with the earth's surface, and on the length of the day. The greatest amount of solar radiation therefore reaches the fringes of the atmosphere over each pole at midsummer, when the

earth's axis is tilted toward the sun, so that radiation hits the earth at a high angle. But because the earth is nearer the sun—roughly 91 million miles—at midsummer in the Southern Hemisphere than it is at midsummer in the Northern Hemisphere—roughly 94 million miles—slightly more solar radiation is received over the South Pole than over the North Pole. (This difference is however cancelled out by the lower temperatures of the antarctic winter.)

The amount of solar energy received at the fringes of the earth's atmosphere at right angles to the sun's rays and at the mean distance from the sun is known as the *solar constant*. According to recent rocket and satellite measurements, the value is almost exactly two *calories* per square centimetre per minute. (A calorie is the amount of heat required to raise the temperature of one gram of water by 1°c.) This is roughly

Right, a diagram showing the wave-length and frequency of long- and short-wave radiation. Wave-length is the distance separating successive wave crests; frequency is the number of wave crests moving past a fixed point each second.

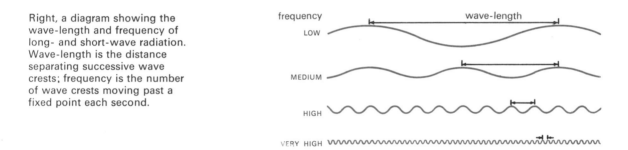

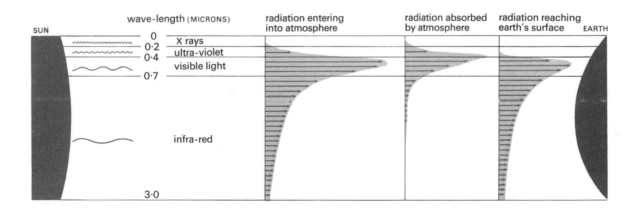

Above, a break-down of the sun's radiation spectrum into the proportions of X ray, ultra-violet, visible light, and infra-red received and absorbed by the earth's atmosphere.

The length of the arrows is proportional to the amount of energy carried by each waveband of radiation. Most solar radiation is in the form of visible light.

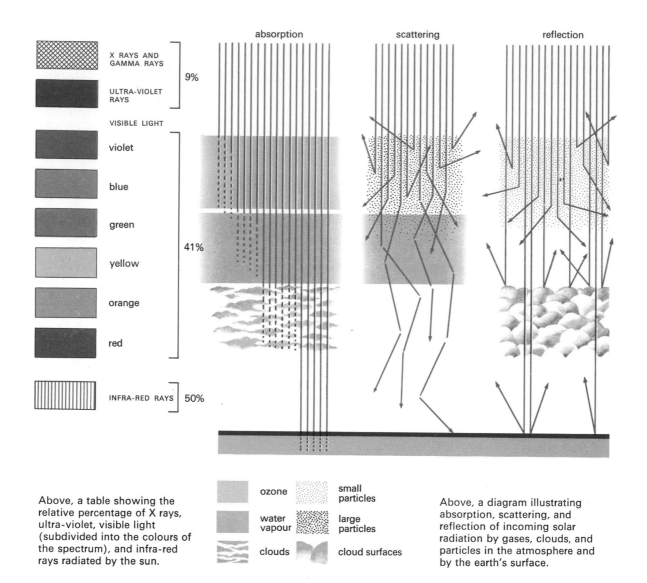

absorption    scattering    reflection

X RAYS AND GAMMA RAYS
ULTRA-VIOLET RAYS
9%

VISIBLE LIGHT
violet
blue
green
yellow
orange
red
41%

INFRA-RED RAYS    50%

ozone    small particles
water vapour    large particles
clouds    cloud surfaces

Above, a table showing the relative percentage of X rays, ultra-violet, visible light (subdivided into the colours of the spectrum), and infra-red rays radiated by the sun.

Above, a diagram illustrating absorption, scattering, and reflection of incoming solar radiation by gases, clouds, and particles in the atmosphere and by the earth's surface.

equivalent to four and a half million horse-power for every square mile of atmosphere at right angles to the sun's rays; or, as the British geographer David Linton pointed out, for the earth as a whole, an amount equal to the total energy output of 180 million very large power stations—so many, in fact, that all the world's land masses would be too small to accommodate them.

The supply of solar energy to the atmosphere is therefore colossal. Yet less than half in fact reaches the earth. We saw in Chapter 2 how ultra-violet radiation at high levels is absorbed as it converts oxygen into ozone; as a result, hardly any ultra-violet solar radiation reaches the earth. Without this abstraction of ultra-violet radiation, life on this planet would probably have evolved differently; and if the ozone layer disappeared today, we should be badly sunburnt and perhaps even blinded. Even so, only a very small proportion of solar radiation is involved in the formation of ozone. Most solar energy is hardly affected by its journey through the atmosphere until it reaches the dense lower troposphere with its concentrations of water vapour, clouds, and dust particles. Here, the sun's rays are strongly absorbed, scattered, and reflected.

Carbon dioxide, oxygen, and water vapour absorb about 9 per cent of the solar energy that reaches the troposphere, and a further

Fresh snow reflects about 90 per cent of solar radiation and so remains firm enough for skiing even in bright sunshine.

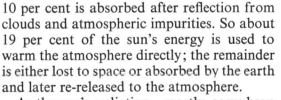

The deep blue of an African sky is caused by the scattering of only the very short light waves; the London sky is greyer because longer waves are also scattered by water vapour and pollution. Scattering is most intense at sunset and sunrise, when only the long yellow and red waves reach our eyes directly.

10 per cent is absorbed after reflection from clouds and atmospheric impurities. So about 19 per cent of the sun's energy is used to warm the atmosphere directly; the remainder is either lost to space or absorbed by the earth and later re-released to the atmosphere.

As the sun's radiation—mostly, as we have seen, in the form of light—passes through the lower atmosphere, it is scattered by any particles that are roughly the same size as the wave-length of the radiation. This selective scattering results in the blueness of the sky, since the short, blue wave-lengths of the light spectrum are obviously more easily scattered than the longer, red wave-lengths. But when the air is contaminated with large

numbers of small motes, such as smoke particles, then slightly longer light rays are also scattered, and the sky becomes a lighter colour, more grey than blue. (In the same way, smoke from a pipe or cigarette appears blue even though the individual particles are coloured brown; this is because the several million nuclei in each puff of smoke scatter the short, blue wave-lengths. After the smoke is inhaled, each particle is coated in a thin film of moisture, which, by scattering more of the longer waves, makes the tobacco smoke appear grey.) This is why there are deep blue skies in many arid lands but rather greyer skies (even when there are no clouds) in the industrial areas of Europe and North

America. Robbed of its blue light, the solar beam now consists mostly of the longer wave-length red and orange light, which passes through the atmosphere without much scattering. This explains why shortly after dawn or before sunset (when the sun's rays make a long traverse of the humid, polluted parts of the lower atmosphere), skies are light blue and the sun's disc, weakened in intensity, appears orange or, nearer the horizon, red. When the sun is low enough in the sky, this light is reflected off the base of the clouds to give the kind of sunset and sunrise that are a favourite subject of poets, artists, and photographers.

Particles and surfaces that are larger than the wave-lengths of light reflect rather than scatter the incoming radiation—in other words, they change its direction instead of its composition. Clouds and the earth's surface are important reflectors of the short wavelengths of solar radiation. (On the other hand, clouds absorb and re-radiate rather than reflect the longer wave-lengths of terrestrial radiation.) The percentage of radiation reflected by a surface is called its *albedo,* and this varies considerably. Freshly fallen snow, for example, reflects more than 90 per cent of the sun's radiation; snow can therefore remain firm even when exposed to bright sunshine. It disappears more quickly when it has a dirty surface or when small areas are cleared in it. The exposed earth, which has a much smaller albedo, then absorbs the sun's radiant energy, which warms and melts the surrounding snow. The very high albedo of fresh snow also contributes to the intense cold of polar regions.

The upper surfaces of clouds are also strong reflectors of solar energy, although their exact albedos depend on their structure and thickness, and on the size of their droplets. A layer of stratocumulus reflects between 55 and 80 per cent of the sun's energy. Even more is reflected if the cloud is made up mainly of small droplets. Clouds composed of small droplets are likely to be rather younger than those that contain somewhat bigger droplets. As a rule these older clouds therefore allow more sunlight to pass through them without reflection, especially near their edges. If we see such a cloud against the sun—particularly when the sun is setting—it is ringed with a bright margin, commonly called a silver lining.

Surfaces on the earth itself also reflect differing amounts and wave-lengths of solar radiation, and it is this ability to reflect light that determines their brilliance and colour. Common surfaces such as rocks, grass, ploughed fields, and forests have albedos varying from 10 to 20 per cent, although dense pine and equatorial forests have albedos as low as 5 per cent, while the albedo of dry sands can be as high as 30 per cent. The albedo of dark, wet soils is generally about 8 per cent, and of grass about 20 per cent, although some of the solar radiation that it reflects is short-wave, and outside the visible light spectrum.

The albedo of water surfaces depends on the altitude of the sun. When the sun's rays are at an angle of only 5 degrees—that is, near dawn and sunset—about 39 per cent of solar radiation is reflected from a calm water surface. This amount falls to 8 per cent at an angle of 30 degrees and to as little as 4 per cent above 60 degrees; at this angle most of the radiation penetrates and warms the surface waters. So seaside resorts facing the sun and backed by high hills may bask in this reflected sunshine when the early-morning or late-evening sun lies low on the horizon.

There are few maps in existence that show albedo values, although one map has recently been produced for North America based upon measurements taken from aircraft. On a global scale, the greatest problem has been the difficulty of estimating the effect of cloud cover, particularly over the oceans. Today, however, satellites can be used to measure the earth's reflectivity, and recent research seems to suggest that the albedo of the earth (and atmosphere) is about 34 per cent. At midsummer, clouds probably reflect about 30 per cent of the solar radiation in high latitudes and 20 per cent in low latitudes; on the other hand, the earth's surface reflects about 12 per cent of the incoming radiation at latitude 80°, but only 1 per cent in the largely sea-covered areas around 20° when the sun is almost vertically overhead.

To sum up so far: if we imagine the amount of solar energy that reaches the earth's atmosphere as 100 units, these are divided up in the following way. About 19 of them are absorbed as they pass through the air; about 34 are reflected back again to space from the tops of clouds and from the earth's surface; and the remaining 47 units (after some scattering) reach the earth where they are absorbed and converted into heat. So less than half the sun's energy that arrives at the edge of the atmosphere eventually reaches the earth, and only about one fifth warms the atmosphere directly; most of the energy that warms the atmosphere comes indirectly from the heated earth by conduction, convection, and evaporation.

We have seen how, in the absence of clouds, more than 80 per cent of the short-wave energy from the sun travels unimpeded through the atmosphere. This is far from true of the long-wave radiation that is the main form of energy radiated by the earth. Only a small percentage of this terrestrial radiation passes through the atmosphere without being temporarily absorbed, mostly by the water vapour and to a smaller extent by the carbon dioxide. This process warms the gases, which in turn emit radiant energy in two directions: upward into the higher atmosphere, and also into the atmosphere below and finally back to earth. The atmosphere's differential reaction to solar and to terrestrial radiation has often been called its *greenhouse effect*, because the near transparency of gases to short-wave radiation and their opaqueness to long-wave radiation are supposed to resemble those of glass in a greenhouse. This analogy is however misleading, since glasshouses do not primarily work in this way. Experiments show that glass is hardly less transparent to long-wave than it is to short-wave radiation. Greenhouses and cloches are effective because they isolate heated air near the ground from the cooler air above; this prevention of mixing is four or five times more effective than the differential reaction of glass to long- and short-wave radiation. For this reason, the term *atmospheric effect* would seem more accurate than greenhouse effect.

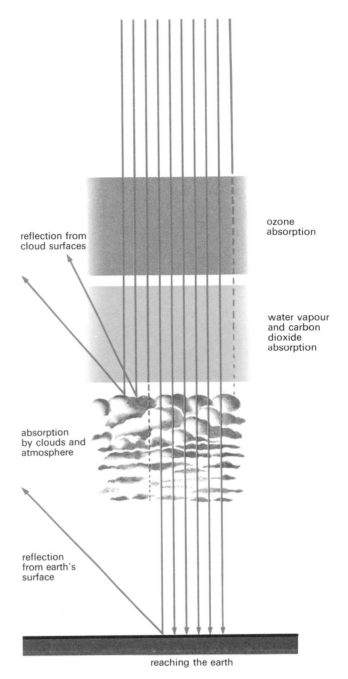

reflection from cloud surfaces

ozone absorption

water vapour and carbon dioxide absorption

absorption by clouds and atmosphere

reflection from earth's surface

reaching the earth

The passage through the atmosphere of 100 units of solar radiation, represented by 10 arrows, each roughly equal to 10 units. About 19 units are either abstracted at high level in the manufacture of ozone or absorbed lower down by water vapour, carbon dioxide, and clouds (after scattering). About 34 units are reflected from cloud surfaces and from the earth's surface. The rest (47 units) reach the earth after some scattering.

At the mean temperature of the earth's surface (about 10°c), terrestrial radiation is within the 4-120 micron waveband—that is, in the form of the long, invisible, infra-red wave-lengths. So there is almost no overlap between the wavebands of solar and terrestrial radiation. We have seen that of 100 units of solar radiation, 47 units were received by the earth. But when we come to analyse the outgoing heat from the earth, we are dealing with a higher figure. The earth's surface is kept warm by the atmospheric effect, so that the number of outgoing units that have to be accounted for is now 119. Of these 119, 105 are re-radiated back to the earth by the water vapour and carbon dioxide of the atmosphere, so that only 14 units are lost to space by direct radiation.

Most of the lost radiation escapes through the "window" that occurs in the water-vapour spectrum between 8.5 and 11 microns—the wave-length at which water vapour fails to absorb radiation. There are also other partially transparent windows in the water-vapour spectrum between 7 and 8.5 microns and between 11 and 14 microns, so that a small amount of terrestrial radiation also slips through these wavebands into space. Carbon dioxide, on the other hand, absorbs strongly in wave-lengths from 12 to 16.3 microns, so any increase in the amount of carbon dioxide in the atmosphere—say, by the burning of carbon fuels—reduces the loss through the partial window in the water-vapour spectrum between 11 and 14 microns. Ozone has a very narrow range of absorption between 9 and 10 microns. But even though terrestrial radiation of this wave-length is strong, there is very little ozone in the atmosphere except at very high levels. So the infra-red absorption by ozone is of only small importance compared with its strong absorption of ultra-violet solar radiation. Similarly, very little terrestrial radiation is absorbed by nitrous oxide and methane.

The atmosphere is warmed not only by radiant energy absorbed from the sun and earth, but also by rising currents of air heated by the earth's surface. These *convection currents* transfer about 10 of the 47 units of solar radiation that reach the earth.

More than twice this amount—23 units of energy—warms the atmosphere by evaporating water from the earth's surface and the release of latent heat (p. 42) into the air as it condenses into clouds. Finally, to balance the amount of energy received, the atmosphere radiates heat to space from its outer fringes by an amount equivalent to the 52 units of energy that it has received directly from the sun and indirectly from the earth.

So if we summarize the various exchanges of energy between the earth, the atmosphere, and space we arrive at the following balancing totals: 100 units of solar radiation reach the outer edges of the atmosphere. Of these,

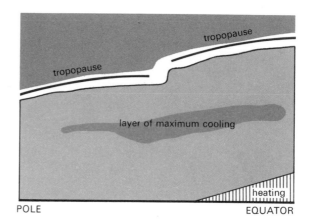

POLE                         EQUATOR

Above, the areas of net radiation heating and cooling in the troposphere. It is the unequal heating of these areas that drives the atmosphere; the winds help to restore the energy balance by transporting sensible and latent heat from low to high latitudes.

In some wave-lengths, the water vapour in the atmosphere absorbs all terrestrial radiation, in others it absorbs part, and in one waveband it absorbs none (see text). This diagram shows the amount of radiation at latitude 50°N; upper curve is for the earth's surface, lower is for the top of the troposphere. Shaded area thus represents total radiant energy lost to space from the earth, and unshaded area is radiation absorbed by the water vapour in the atmosphere.

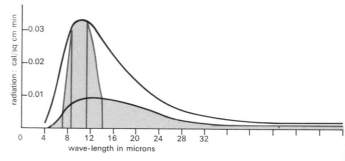

34 are instantly reflected back into space from clouds and the earth's surface; 14 are lost to space by direct terrestrial radiation; and the 52 that are absorbed by the atmosphere are finally radiated back into space. Of the 52 units that are taken into the atmosphere, 19 are absorbed from solar radiation; 10 warm the air by convection and 23 by evaporation and condensation. As for the 47 units of solar radiation that reach the earth after some scattering, 14 are lost to space, 23 are used for evaporation, and 10 for convection. In this way, an overall balance is struck between the energy receipt and loss in all three parts of the system.

In broad terms then—that is, allowing for a 10 per cent margin of error—it is possible to balance the earth's heat budget, although at this scale, we still do not know exactly how the two sides of the radiational energy equation—that is, receipt and loss—cancel each other out. In every region, there is almost always a surplus or deficit of radiational energy, either at one particular season or even throughout the year. As we shall see in the next chapter, it is these regional differences that keep the atmosphere on the move, for it is the winds and ocean currents that, by transporting heat in a variety of forms, restore the local balance of energy.

Diagram illustrates the figures quoted in the text concerning solar radiation and terrestrial radiation. It splits the amount of solar radiation into 100 units. Because of the atmospheric effect the amount of terrestrial radiation to be accounted for is 119 units. Minus figures indicate a loss of radiation; plus figures, a gain of radiation.

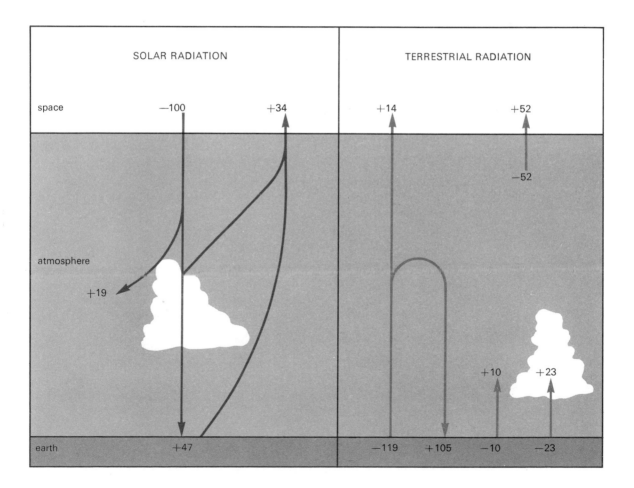

# 4  The Atmosphere in Motion

The atmosphere is perpetually on the move over distances that vary from a few inches to thousands of miles. And as air moves, it transports heat, moisture, and momentum whether it is a small, local breeze travelling from an open park to a city street or a huge wind current flowing from the tropics to higher latitudes. In this chapter we shall be considering mainly the large-scale transports of the general circulation rather than the exchanges involved in local winds. But almost all winds, on whatever scale, not only carry energy but are also produced by it: they are the result of the unequal heating of the earth's surface by the sun.

To understand how the varying impact of the sun at different latitudes and on land and sea sets the atmosphere in motion, we must first explain the different forms of atmospheric energy and their interchange from one form to another. Obviously the initial effect of heat energy from the sun is to raise the temperature of both the atmosphere and the land and sea beneath it. As soon as this happens, some of the original solar energy begins to be converted into other forms. At first the energy is *internal*—that is, the energy of molecular motion, which we generally know as heat. Then, because air has been heated, it expands so that its centre of gravity is raised, thereby increasing its potential energy—called *geopotential* energy in meteorology. The combination of heat energy and geopotential energy is known as *total potential* energy. Yet another form of energy is contained in water vapour, which is produced by evaporation from the sea's surface, from cloud droplets, from moist ground or rivers and lakes, or from the surface of leaf tissues—called transpiration. We call this *latent* energy, since the conversion of water into vapour demands a great deal of heat, called latent heat, which is again given up when the vapour condenses

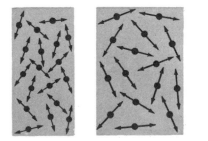 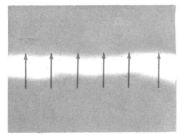

Diagrams representing the four types of atmospheric energy, all of which derive from the sun.

*Internal* energy is the energy or heat of moving air molecules, and is measured as temperature. (Air on right is hotter than left.)

When air is heated, it rises; its centre of gravity is therefore raised, and its *geopotential* energy is increased.

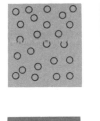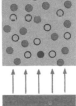

Water vapour also contains latent energy; this is the *latent heat* used to evaporate water that is released again on condensation.

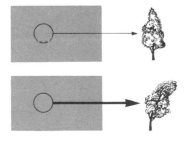

Part of the atmosphere's total energy is finally converted into the *kinetic* energy of motion of the winds.

To a meteorologist, the wind-filled sails of this yacht are evidence of air in motion. Air moves in response to inequalities of heating set up by the varying impact of solar radiation on the earth's surface; its motion also helps to balance these inequalities, by transporting heat from the tropics to higher latitudes.

back into water. Finally, the motion of the air—caused originally by its intake of heat or internal energy—gives it energy of motion, or *kinetic* energy.

The atmosphere is constantly converting energy from one form to another but—except when there are gains or losses from radiation—the total remains constant. The process is as follows: solar radiation absorbed by the earth and atmosphere is converted into two forms of energy—internal and geopotential. This total potential energy is then changed by convective processes partly into the kinetic energy of the winds and partly into the latent heat of water vapour. Both types of energy are later reconverted into heat: the kinetic by friction set up between the wind and the earth's surface and between the molecules of air, and the latent by condensation.

Most of the atmosphere's energy is internal and potential: only a small proportion is kinetic and latent. And because the total amount of kinetic energy of the atmosphere remains almost constant, it follows that it is generated at the same rate as it is dissipated. Expressed in joules, or units of work, the amount of atmospheric kinetic energy has been estimated at about two joules per square metre—a figure that represents only about 0.57 per cent of the sun's energy entering the atmosphere. (One joule is equal to 0.738 foot-pound per second or 0.24 calories per second.) So even if we take only the amount of incoming solar radiation that is actually used to heat the earth and atmosphere—in other words, if we ignore the 34 per cent that is lost by reflection—the energy of motion is still less than one per cent (0.95 per cent). Even so, this tiny

In the first 3000 feet or so of the atmosphere, winds are affected by friction with the earth's surface. Frictional drag not only reduces the speed of a surface wind but also alters its direction. Over a smooth, level surface, such as the featureless polar region (left), the wind direction may differ from that at 3000 feet by 10° and its speed may be cut by about a quarter. Over rougher country, such as the Welsh hills (right), the surface wind may travel at half its speed at 3000 feet and the difference in direction may be as much as 25°.

percentage is sufficient to keep the atmosphere active enough to balance out the unequal distribution of heat over the earth produced by the sun's and earth's radiation.

The relatively small amount of kinetic energy in the winds can be illustrated in another way. The American meteorologist H. H. Lettau has estimated the total kinetic energy of the winds at 140 watt-hours per square metre (that is, at the rate of one joule per second), and the average rate of dissipation of energy by the winds at two watts per square metre. From these figures, we might conclude that in 70 hours the total kinetic energy of the winds would be dissipated. We should however be forgetting that the speed of dissipation would decline as the available kinetic energy was reduced. Another American meteorologist, S. L. Hess, allowing for this decrease in speed, estimated the rate

of dissipation at about 36 per cent per day. It would therefore take just under 13 days to reduce the atmosphere's reserve of kinetic energy by 99 per cent.

In practice, of course, the wind's energy is *not* dissipated at an even rate over the earth's surface, for two reasons. First, the amount of frictional drag over the smooth, level surfaces of the great polar ice-caps or the calm waters of lakes and seas is far less than it is over the rough, uneven surfaces of most land areas. Second, the rate of dissipation is proportional to the winds' velocity, so much of the kinetic energy of the winds is lost, or rather converted by friction into heat, in deep cyclonic storms.

Having seen how kinetic energy is dissipated, the next question that arises is how it is produced from other forms of energy. Some 60 or so years ago, Max Margules, an

The general pattern of winds over the earth.

In low and high latitudes in each hemisphere, air circulates through sloping cells known respectively as the Polar Cell and the Hadley Cell, the latter being much bigger and stronger than the former. In middle latitudes, air moves through a series of gigantic waves, rising as it moves poleward in one limb and sinking as it moves equatorward. This simple pattern of the general circulation excludes, of course, all day-to-day complications of air movement.

Austrian meteorologist, showed that kinetic energy is generated by a convective system linking rising warm air with sinking cold air. In such conditions, the heavier, cooler air lowers the system's centre of gravity and reduces its geopotential energy. Unless translated into other forms of energy, the amount of kinetic energy therefore increases in proportion to the decrease in geopotential. But if, on the other hand, warm air is forced to sink and cold air to rise—and so raise the centre of gravity and increase the amount of geopotential energy—then the kinetic energy of the system is reduced.

The atmosphere as a whole carries a great deal of energy in the various forms we have described, but not all of it is convertible into kinetic energy. What we have to consider is the proportion of potential energy (amounting in fact to less than 0.5 per cent) that is able to convert into kinetic energy and so produce air movements. In other words, we have to look at the atmosphere's reserve of *available* potential energy—specifically, energy that is produced by inequalities of heating. These may be inequalities of heating between one latitude and another, or from one longitude to another, and may usefully be distinguished as *zonal available potential energy*—produced by contrasts in temperature between the tropics and poles—and *eddy available potential energy*—mainly produced by the unequal warming of land and sea. It has been calculated that the zonal form of potential energy is about ten times greater than the eddy type. Even so, the eddy form is still very important in the generation of kinetic energy, since zonal available potential energy is converted into eddy available potential energy by a process known as the *eddy heat flux*. This occurs when warm southerly airstreams alternate with cool, northerly airstreams as in latitudinal waves or a series of depressions and anticyclones.

In the atmosphere, the generation of kinetic energy takes place on a large and a small scale. Minor local storms generate little kinetic energy, and even the considerable amount created by a tropical cyclone is quickly dissipated in overcoming the surface friction between the strong winds and the ground. So much so that the British geographer F. K. Hare has estimated that without loss caused by friction the energy carried by no more than 20 tropical cyclones would be enough to drive the entire tropospheric circulation.

A major supply of kinetic energy in the atmosphere is generated by the gigantic overturnings of air in the tropics. In tropical latitudes, there is a near-surface flow of cool air toward the equator; these winds are known as the *trades*. At the equator, the air rises and returns toward the poles at higher levels, reaching maximum speeds at about 38,000 feet. This circular movement of air rising near the equator and sinking near the tropics is called the *Hadley* (or *trade wind*) *Cell*. As they move toward the equator over the sun-warmed sea, the trades pick up both heat and moisture by evaporation, and are therefore loaded with water vapour containing latent heat. When condensation into clouds takes place, this heat is reconverted into what is known as *sensible* heat, which warms the atmosphere. This combination of sensible and latent heat provides enough kinetic energy both to overcome frictional losses and to carry a considerable balance of energy back to mid-latitudes in the form of high-level eddies or waves. This feed-back of kinetic energy may be even greater in the Southern than in the Northern Hemisphere. Certainly, it helps to carry the powerful westerly current of air at about 38,000 feet (known as the *sub-tropical jetstream*) that is found in both hemispheres.

Meteorologists once thought that there was another overturning or cell in intermediate latitudes. This was known as a *reverse* cell, since it was thought to consist of

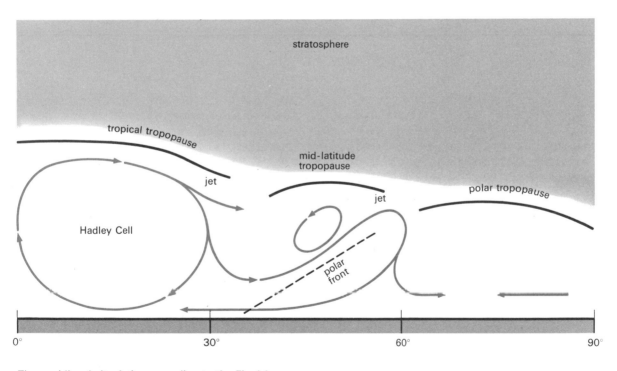

The meridional circulation according to the Finnish meteorologist E. Palmén. It shows the circular movement of air in the tropics known as the Hadley Cell and also the high westerly sub-tropical jetstream. The diagram also sets out Palmén's theory of a slanted current of rising warm air at the polar front. As it travels poleward, this air sinks and moves horizontally north and south.

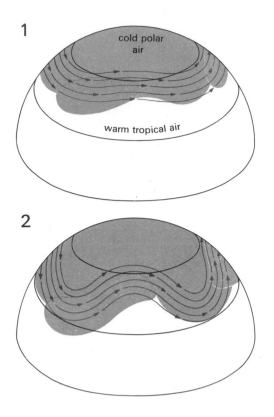

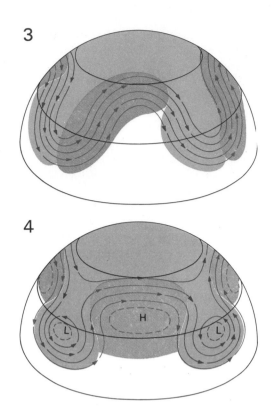

air sinking near the tropics, moving poleward and then rising in higher latitudes—that is, rotating in the opposite way to the Hadley Cell. Recent research however seems to show that the only major meridional (south-north) cell is the Hadley Cell. According to the widely accepted theory worked out by the Finnish meteorologist E. Palmén, there is, in place of a reverse cell, a slanted convection current of rising warm tropical air in waves in the *westerly winds* of mid-latitudes. This warm air flows over cold polar air at the so-called *polar front* (which is actually in middle latitudes), then sinks in higher latitudes to disperse north and south in horizontal eddies near the ground. In subpolar areas there is, according to Palmén, a zone of horizontal mixing of different airstreams. But though few modern meteorologists accept the old idea of a three-cell circulation with a distinct reverse cell, many still believe there to be a minor overturning in the mid-latitude westerlies. This system is known as the *mid-latitude* or *Ferrel Cell,* and

its most important and distinctive characteristic is that it forces warm air to sink (in low latitudes) and cool air to rise (in higher latitudes). As a result, kinetic energy is reconverted into zonal potential energy; even so, the process is thought to be slow and intermittent. Lastly, in very high latitudes, there is known to be a very shallow *Polar Cell,* consisting of sinking air above the poles, and rising air in slightly lower latitudes. The circulation is weak, however, and little kinetic energy is generated compared with that produced by the Hadley Cell.

A great deal of kinetic energy is also produced and latent heat transported by the mainly horizontal airstreams of the troposphere and lower stratosphere in middle and high latitudes. The most helpful way to think of these is in terms of size. Far the largest are the two broad currents of high westerly winds that flow around each hemisphere. These winds travel through a series of huge, slowly moving, or for long periods almost stationary, waves between high and

Left, stages in the development of high-level waves in the Northern Hemisphere. First, the boundary between the cold, dry polar air (blue) and warm, moist tropical air (pink)—the path of the fast-moving westerly jetstream—begins to undulate. As waves form, polar air is pushed southward and tropical air moves northward. Finally, the waves detach themselves into isolated cells of warm (high-pressure) air and cold (low-pressure) air.

Right, a map centred on the North Pole showing the interlocking pattern of high- and low-pressure areas associated with the development of high-level waves.

low latitudes, particularly in the Northern Hemisphere. The great mountain ranges of the Western Cordillera (including the Rocky Mountains and the Sierra Nevada), the Andes, and the high plateau of Asia help to control the position of these gigantic ridges and troughs, and there are normally from two to four waves around each hemisphere. Warm ridges of high pressure are commonly found high over the eastern Pacific and western North America and over central Asia, and low pressure troughs of cool air over eastern North America and the western Atlantic and over eastern Asia and the western Pacific. The form and position of the waves also varies with the speed at which the air is moving through them, so there are periodic shifts in the general pattern of pressure and in the flow of the winds themselves, which in turn affect the broad character of the weather.

Smaller than the large, almost stationary, long waves in the upper westerly winds, but very important as generators of kinetic energy, are the largest of the travelling waves in middle latitudes. They travel through the larger waves we have just described to produce a complex pattern of circulation at intermediate heights. At the ground, the ridges of the waves are represented by anticyclones and the troughs by large depressions. Many meteorologists once viewed these moving mid-latitude disturbances as little more than turbulent incidentals in the broad flow of the general circulation. Since World War II, however, several meteorologists have helped to establish the vital part played by the travelling wave depressions in producing kinetic energy and in distributing heat, water vapour, and momentum.

In the middle and upper troposphere, the disturbances are in the form of a series of steep waves with a wave-length of about 60° of longitude. As air moves around these tight ridges and troughs, it alternately converges and diverges—a movement that often leads to a compensating divergence and convergence in the air below and helps to produce anticyclones and depressions near the surface. In this way, these high-level waves are responsible for much of the day-to-day weather in mid-latitudes. With their wave-like form at high levels and more circular pattern nearer the ground, these travelling disturbances generate a great deal of kinetic energy as cool air sinks behind the trough line of the wave—that is, in the rear of the depression—and warm air rises toward its crest—that is, in the front of the disturbance. It would seem in fact, that in the Northern Hemisphere two or three of these travelling wave depressions would generate enough kinetic energy to drive the whole circulation north of 30°.

To sum up so far: the troposphere is mainly driven by the kinetic energy generated in the Hadley Cell overturnings of low latitudes and the moving wave depressions of mid-latitudes. Some of the kinetic energy created in the Hadley Cell is carried to middle latitudes by large-scale eddies, mainly in the upper troposphere. Together, these two sources of kinetic energy are powerful enough to balance losses of energy through friction and in any Ferrel (reverse) overturning. Above 65,000 feet, the atmosphere seems to be set in motion by a new, independent source of available potential energy, produced by the unequal heating of the ozone layer by the sun.

These then are the main ways in which potential energy is converted into the kinetic energy of the winds. But before we explain how these movements of air are able to balance and redistribute heat, water vapour, and momentum, we need to look a little more closely at the overall pattern of the general circulation and the major disturbances associated with this pattern.

As we have already seen, the atmosphere is set in motion by the unequal heating of high and low latitudes and of different surfaces by the sun. Variations in heating create areas of different pressure, which falls more gradually with height in warmer, lighter air than it does in colder, heavier air. For this reason, pressure differences are created above the earth, and in the absence of other complicating factors, winds blow

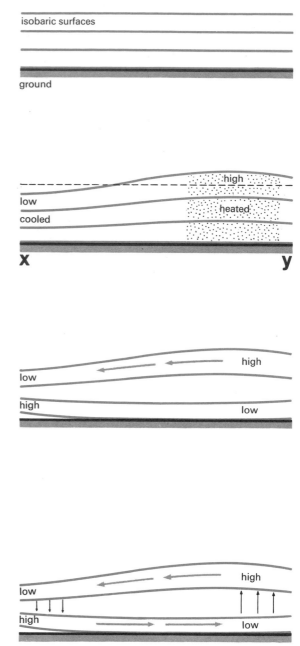

Four diagrams illustrating a simple convective system of winds produced by the unequal warming and cooling of the lower atmosphere. First, imagine an air layer of uniform temperature and pressure. Next, the air above Y is heated and so expands, and the air above X is cooled and so contracts. Pressure is now higher above Y than it is at the same level (shown by the dotted line) above X. Air now flows from warm to cool along this pressure gradient at high level. This transfer of air increases the pressure at X and reduces it at Y. Air therefore moves along the pressure gradient from X (cool) to Y (warm) at low level. The system is completed by a general rising of air at Y and sinking at X.

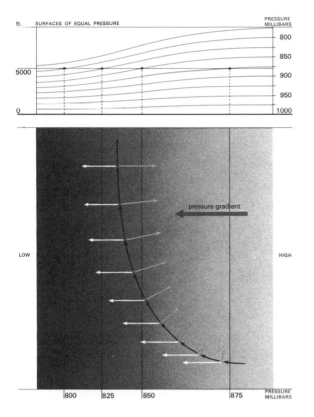

from the warmer to cooler air at high levels and from the cooler air to the warmer at low levels. The flow of circulation is completed by rising currents in the warm air and descending currents in the cool air. Outside equatorial regions, however, several other influences combine to complicate the large-scale movements of air.

A major influence on the pattern of winds is the effect of the earth's west-east rotation, which generates an apparently deflecting force known as the *Coriolis force* (or sometimes the *geostrophic force*). This acts to the right of a moving wind—in relation to the direction in which it is travelling—in the Northern Hemisphere, and to the left in the Southern. As a result, air that is moving initially from an area of high pressure to an area of low—that is, down the pressure gradient—is gradually deflected more and more to one side by the Coriolis effect. But the effect ceases—if no other forces (such as surface friction) are operating—as soon as air

A diagram showing the influence of the Coriolis effect (see below) in the Northern Hemisphere. As soon as air begins to move down a pressure gradient (seen in section at the top of the page), it is increasingly deflected to the right until a balance is struck between the Coriolis effect and the pressure gradient; the air then flows along the isobars with low pressure to the left.

→ centrifugal force

→ Coriolis force

→ pressure-gradient force

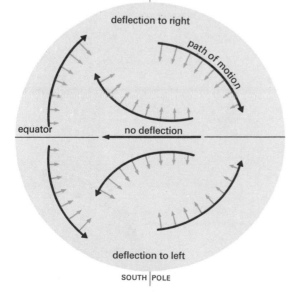

A diagram illustrating the deflective effect of the Coriolis force (which results from the earth's west-east rotation). This so-called force deflects any moving body to the right of its path in the Northern Hemisphere and to the left in the Southern.

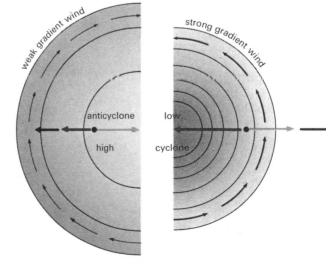

Above: diagram representing the balance of forces in an anticyclone (left) and a cyclone (right) in the Northern Hemisphere. The nature of the balance in the anticyclone prohibits intense pressure gradients and very strong winds.

begins to flow at right angles to the pressure gradient, that is, parallel to the isobars. The result of this balance—called the *geostrophic balance*—between the Coriolis force and the pressure gradient is a steady wind flowing parallel to the isobars.

When isobars are tightly curved, as, for example, near the centre of depressions and particularly in tropical storms and tornadoes, the flow of air is affected by a third factor: *centrifugal force*. This tends to fling the air outward from the centre of rotation—with a force that is proportional to the square of the wind velocity as expressed by the steepness of the pressure gradient and inversely proportional to the radius of rotation. But in many of the largest anticyclones, pressure gradients are gradual and the isobars are only very slightly curved; winds are therefore light, and near the centre they may be too gentle to measure accurately.

In the lowest 3000 feet or so of the atmosphere, winds are influenced in yet another way, this time by the force of friction between the moving air and the earth's surface. Frictional drag slows the wind down and causes it to blow across the isobars toward an area of low pressure. This reduction in speed and change in direction varies with the roughness of the land: over very uneven ground, the speed of the wind can be cut down by friction to as little as half its speed at 3000 feet; it then blows across the isobars at an angle of up to 25°.

The fickleness of the weather is notorious and its caprices are daily confirmed by the

A cross-section of the troposphere showing the prevailing global winds from pole to pole.

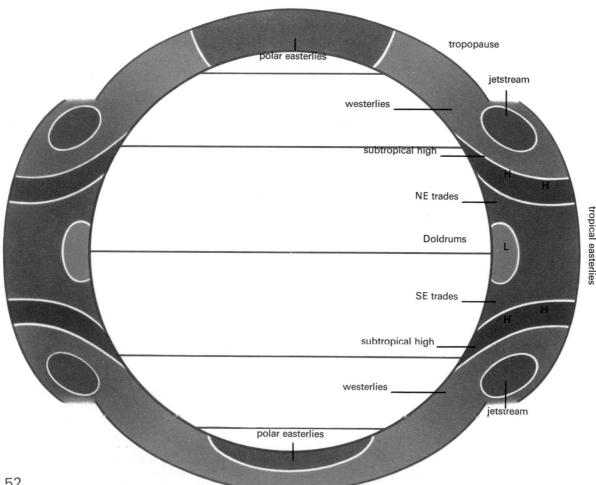

weather charts that appear in newspapers and on television. Yet if averages of wind speed, temperature, and so on were charted over a single season or a whole year, the more ephemeral features such as travelling depressions and anticyclones would filter away, and a simpler, bolder pattern of winds would emerge. On such a chart, we should be able to recognize several broad areas in which the atmosphere moved in a reasonably distinctive and uniform way. So let us for the moment ignore all complicating factors, and take a look at the broad belts of pressure and wind that make up the main features of the general circulation.

Near the equator there is a belt of low pressure—called the *Doldrums* by the captains of early sailing ships—which for most of the year produces calms or light winds. At the equinoxes however (when the sun is vertical at noon over the equator) thundery squalls are more common. These squalls form on the *intertropical convergence* (or *front*) where the trade winds from the two hemispheres are drawn together. This zone of convergence moves with the seasons, with a time lag of up to one month behind the vertical sun as it migrates north and south. (It also travels farther over land than sea because land warms faster than water.) On either side of the Doldrums are the areas of high pressure known as the *Horse Latitudes*. Winds move outward from these areas, and on their equatorial sides the trade winds blow from the north-east in the Northern Hemisphere and from the south-east in the Southern. The wind direction is consistent only in the Atlantic Basin, however, where it is unaffected by the monsoons of south-east Asia. Beyond the Horse Latitudes—between 35 and 60° in both hemispheres—pressure falls poleward and there is a zone dominated by westerly winds. These are more variable than the trades, especially in the Northern Hemisphere, where they blow around a constantly changing sequence of depressions and anticyclones. Lastly, in very

One of the many factors that disturb the pattern of winds is the effect of the earth's high mountain ranges. This view depicts the Rocky Mountains, which, with the Andes and the Asian plateau, partly control the position of the huge waves through which the upper westerly winds travel around each hemisphere.

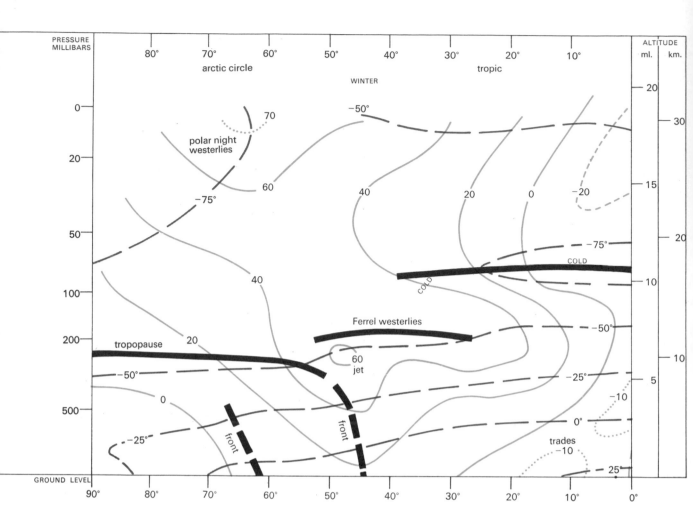

Above, two-part cross-section relating temperature and wind speed in the first 20 miles or so of the atmosphere near longitude 30° W in January. The broken grey lines indicate the temperature levels in °C. Wind speeds (in knots) are shown by blue lines in terms of *west* wind speed; the speed of the westerlies is therefore represented by positive values and that of the easterlies by negative ones.

high latitudes, there are the shallow high-pressure areas over the poles with out-blowing winds.

This then is the simple outline of the earth's surface winds. Its pattern changes radically with the seasons, however, and with the unequal heating of land and sea by the sun. For as the vertical sun appears to move north and south with the seasons, the wind systems following in its path change in location, intensity, and even in basic form. For example, mid-latitude depressions are more common and occur nearest to the equator during the winter. Sub-tropical anticyclones, on the other hand, expand and move toward the poles in summer. Over the continents, temperatures rise and atmospheric pressure falls in summer, while in

Right, a drawing showing the tube-like structure of a jetstream. The fast-moving core of air, which can vary in length from a few hundred to several thousand miles, travels through the waves in the upper westerlies at a height of about 38,000 feet.

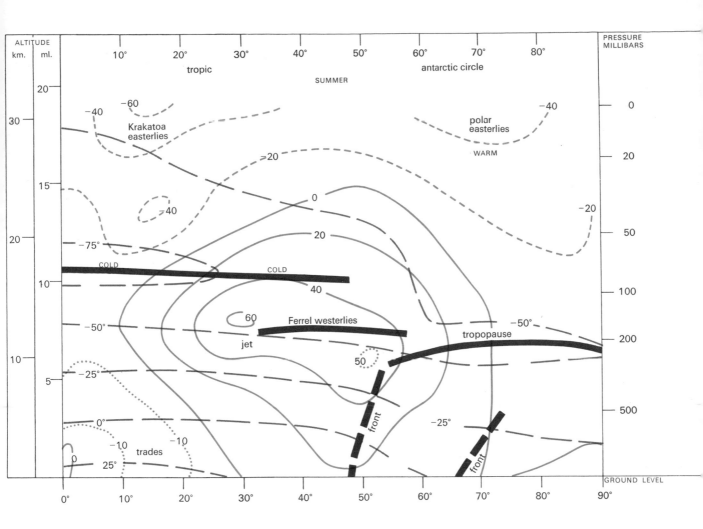

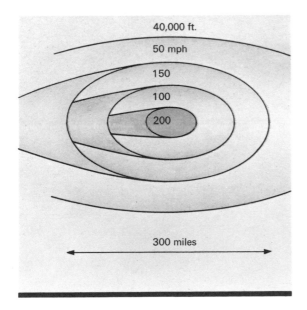

winter the opposite is true. This seasonal change is most clearly seen in the monsoonal reversal of pressure and winds over Asia.

The tidy pattern of winds is also disturbed by the world's highest mountain belts—in particular, the Rocky Mountains, the Andes, and the Plateau of Tibet. As we saw earlier, these upland areas help to throw the upper westerlies into a series of waves between high and medium latitudes. And since these high waves relate to the position and form of the major anticyclones and families of depressions near the ground, they are partly responsible for seasonal changes of pressure over continents and seas.

To get some idea of the seasonal variation in the general circulation, it is worth looking in turn at the general pattern of winds in the

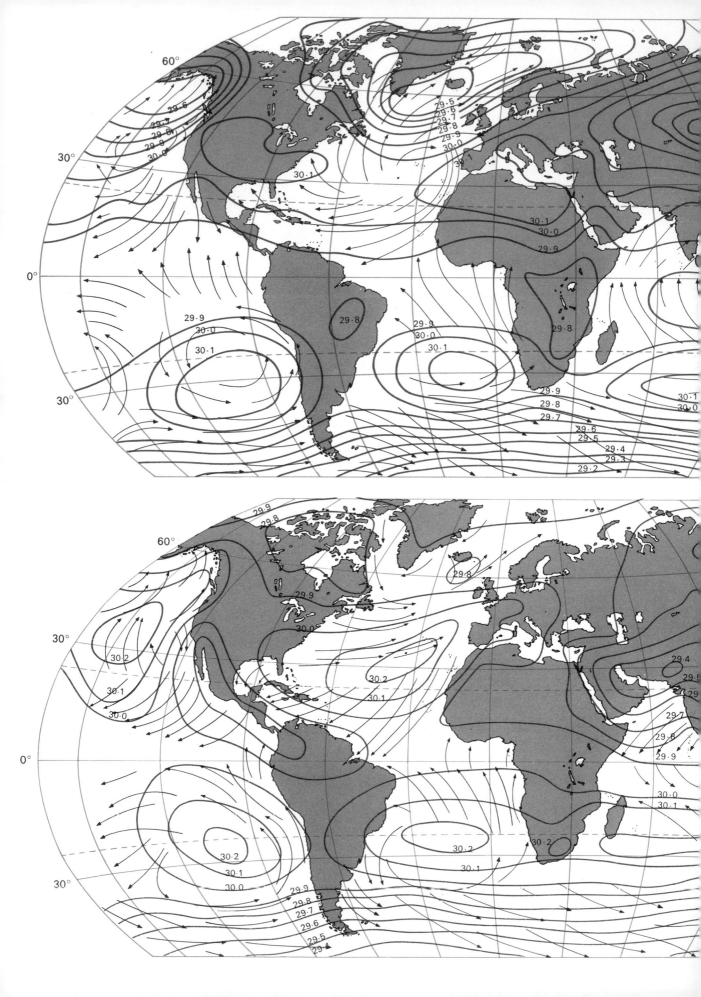

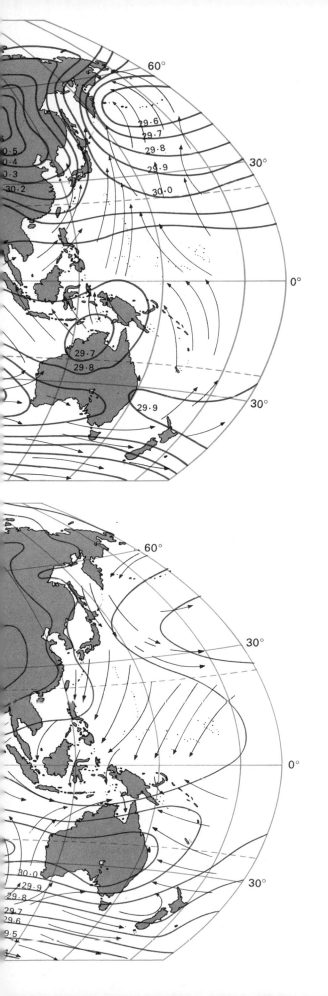

"summer" and "winter" hemispheres—that is, in either hemisphere during summer and winter. When taken in section, the summer hemisphere presents a simple picture. Below about 12 miles in tropical latitudes, there are the light and somewhat variable easterly trade winds; at higher levels, the winds increase in speed to more than 60 knots. These are known as the *Krakatoa easterlies,* because they carried the dust from the volcanic eruption of the Indonesian island of Krakatoa in 1883 around and around the earth for more than three years—and, incidentally, caused spectacular sunsets. Over the poles, these winds link up with the rather weaker easterlies known as the *polar easterlies.*

Between the low- and high-latitude belts of easterly winds is the great westerly system of winds known as the *Ferrel westerlies.* Blowing between the surface and a height of about 15 miles, these winds increase in strength to form two or three very fast-moving cores within the wind system. These are known as *jetstreams,* and lying at heights of about six to seven miles (at or just below breaks in the tropopause) their central speeds vary from 60 to 200 knots or more. Jetstreams were first encountered by high-flying aircraft during World War II, and since then these westerly air currents have been charted by radiosondes, aircraft, and rockets. Today aircraft take advantage of them on west-east flights. (There is only one substantial *easterly* jetstream, which develops in summer high above the Indian Ocean in the Northern Hemisphere.) These rivers of fast-moving air vary in length from a few hundred to several thousand miles; at either end, there are cross-jet movements, with rising currents to

Two maps showing the pattern of the general circulation during January and July. The winds are represented by black arrows and the pressure systems by red isobars.

57

the right of the entrance and to the left of the exit that seem to have some connection with areas of deepening depressions.

In winter, the positions and intensities of the various wind systems are very different. In high latitudes (above about 65°) the easterly winds are much shallower than they are in summer. And at heights of more than 10 miles, the easterlies are replaced by a strong westerly vortex around the pole known as the *polar night westerlies,* with a jetstream core and winds travelling at more than 200 knots at heights varying from 15 to 20 miles between latitudes 55 and 65°. In midwinter, the Ferrel westerlies, which carry fast-moving and generally intense depressions and intervening ridges and anticyclones in the lower atmosphere, are strongest and most widespread. In summer, they extend generally from about 35 to 65° at sea level, but in winter from 30 to 70°. (At about 18,000 feet, they extend from much nearer the equator to the pole.) In the lower stratosphere, the winds decline rapidly in speed to an average of 25 knots at about 15 miles. Higher still, in the upper stratosphere, the winds again pick up speed to reach a fast-moving core at about 35 miles—that is, in the lower mesosphere. These high-level, strong westerly winds are known as the *mesospheric westerly current.*

All these air currents contain travelling disturbances. The mid- and high-latitude westerlies in particular are characterized by moving depressions and anticyclones in the lower troposphere and, as we saw earlier, by waves of varying dimensions in the middle and upper troposphere and lower stratosphere. In fact, a chart showing the pattern of pressure at about two miles would differ very little from a similar chart for a height of 15 or even 30 miles. Even so, the atmosphere does not behave consistently at all levels. For example, winds in the stratosphere move from cold to warm areas and not the other way around, as is the general rule in the troposphere. At stratospheric levels, the circulation is more of a cooling than a heating system, accentuating the contrasts in temperature over different parts of the earth rather than levelling them out. Above about 50 miles or so, the atmosphere is ionized, and seems to be influenced by yet another set of factors.

A great deal of information has been assembled recently about the general circulation of the atmosphere, particularly in the Northern Hemisphere. But observing is not the same as understanding, and even now it is difficult to relate our present knowledge of the atmosphere to the climates of the earth. Recently, however, our understanding of the general circulation has been enormously increased by a series of investigations carried out by British, American, and Scandinavian meteorologists. As a starting point, these studies assumed that the total amounts of both energy and water vapour, although unequally distributed, must remain constant throughout the atmosphere. From here, the next step was to discover where these areas of input and output are to be

A diagram representing air moving from the equator (A) to the North Pole and from high latitudes (B) toward the equator. As it nears the earth's axis of rotation, the north-bound air travels progressively faster; conversely, the south-bound air travels more and more slowly. This is because the atmosphere possesses angular momentum, which is the product of mass × velocity × radius, and is constant for any body moving around a fixed axis. So, in the absence of any other force, any decrease in radius results in an increase in speed, and vice versa.

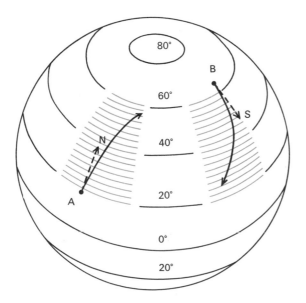

found and how supplies are exchanged from one to the other.

To understand the workings of this transfer system, it is first necessary to remember that the atmosphere not only rotates with the earth, but also moves on its own around the earth's axis. In other words, the atmosphere possesses *angular momentum*. The angular momentum of a body moving in a circle is proportional to its velocity and to its distance from the centre of the circle (in this case, the earth's axis). Angular momentum is in fact a product of three factors—mass × velocity × radius. We have seen how low latitudes are generally dominated by easterly winds (the trades) and middle latitudes by westerly winds (the westerlies). Because of the friction between these winds and the earth rotating beneath them from west to east, westerly angular momentum is generated in low latitudes, which is therefore a source, and is transferred to sink areas in middle latitudes. The effect of this surface drag would be to bring both circulations to a standstill in about 10 days, were there not this transfer of westerly angular momentum from low to high latitudes.

Angular momentum is transferred from low to high latitudes in a number of ways: first, by movements in the Hadley Cell in low latitudes, and second, by travelling disturbances. The second process is most pronounced in the upper troposphere with a maximum transfer at about 30,000 feet in latitude 32°—that is, in the sub-tropical anticyclonic belts. Third, momentum is almost certainly transferred by the great high-level tropospheric pressure waves and their accompanying family of surface cyclonic and anticyclonic disturbances.

Since there is always a limit to the amount of heat gained or lost in different parts of the world, it follows that energy is transferred from low latitudes, and low levels, where there is a surplus amount of energy from the sun, to the much larger areas of net radiational heat losses in high latitudes and high levels. The winds are both an effect of these inequalities and a method of balancing them by the transfer of energy in various forms. The flow of kinetic energy is, as we

have seen, relatively small compared with other forms of energy exchange. The part played by ocean currents as a means of transferring sensible heat is probably important, but detailed measurements are still needed to confirm this. Between latitudes 30 and 40°, where there is the greatest interchange of energy, the parts played by sensible and latent heat in the transfer of energy are more or less equally important. In higher latitudes, there is a greater transfer of energy as sensible heat in the form of warm air currents. Latent heat is released when water vapour condenses into the tiny water droplets of clouds. This process is most intense at one to four miles in tropical latitudes—at the same level, in fact, that there is the greatest amount of long-wave, radiative cooling.

Like energy, there is a roughly uniform amount of moisture in the atmosphere, in spite of inequalities of evaporation and precipitation in certain areas. And like heat, moisture is carried from source regions— where evaporation exceeds precipitation—to sink areas—where the reverse is true. In this way, a certain local and global balance is achieved. As yet we are unable to calculate the amounts of moisture involved in these transfers, partly because we do not yet know how much precipitation falls over most of the oceans and large areas of the southern continents. Also no satisfactory instrument yet exists for measuring evaporation, although reasonably accurate estimates can usually be made from the values of other elements, as we shall see in Chapter 5. The few regional studies that have so far been made show that sources and sinks of water vapour are not so closely related to latitude as are those of momentum, or even of heat energy. For example, the widely separated Gulf of Mexico and the north-east Pacific provide about 90 per cent of all the precipitation falling over the entire Mississippi watershed. It has also been found, paradoxically, that certain arid regions are also source areas for water vapour. (Here we must assume a convergence of water into these areas by sub-surface percolation or surface run-off.) Not surprisingly, the least amount of evaporation occurs in polar regions because of low temperatures and frequent calms. But in middle latitudes, especially where there are strong winds and warm seas, evaporation is intense. This is particularly true of the western North Atlantic and Pacific Oceans, where as many as 100 inches are evaporated each year.

From what we have said so far, it is obvious that the general circulation is a rather more complex system than it is sometimes represented to be. It is therefore not surprising that there are still some very important questions about its behaviour still to be answered. For example, there is still a lot to learn about the westerly winds that play such an important part in the atmosphere's circulation. Solutions to the problems posed by the general circulation are likely to be found by following three general paths of inquiry: theoretical, experimental, and observational.

In most theoretical investigations, a simplified model of the atmosphere is used to formulate a series of mathematical equations defining the application of certain physical laws and relationships. If these equations are then found to correspond to the way the atmosphere's circulation actually behaves, they may then lead to valuable new discoveries. Such experiments were on the whole impossible before the development of modern electronic computers. The American meteorologist N. A. Phillips, for example, has used a computer to construct a very realistic model of the atmospheric circulation, with which he was able to show that travelling depressions and anticyclones have a considerable influence on the development of strong westerly winds outside the tropics.

In the laboratory, experiments are usually made with a "dish-pan" apparatus. This is either a hemisphere or a concentric cylinder filled with fluid, which is rotated at various speeds while its core (representing the pole) or perimeter (equator) is cooled or heated. In this way, D. Fultz, of the University of Chicago, successfully simulated the distribution of heat in the atmosphere around a rotating earth. An important result of these experiments has been to stimulate ideas that could later be verified by observation.

Few of the mathematical, physical, and experimental models of the atmosphere that have so far been used take into account the distribution of land and sea or the gigantic monsoonal wind reversals—topics that are often the subject of fascinating and controversial studies. In fact, no reconstruction of the atmospheric circulation, whether on paper or in the laboratory, can do more than give an idea of its general character. It does not follow, of course, that investigations in this field are unimportant: they are vital if we are to understand how the atmosphere ticks. But by definition, any picture that deals only in averages filters out the small-scale features that may be vitally important in local weather. The next chapter will take a rather closer look at some of the processes that produce day-to-day weather rather than seasonal climates.

Above, whirling currents in a "dish-pan" experiment at the University of Chicago to simulate the major air currents. The pan's rim (representing the equator) is heated, and the centre (the pole) is cooled. Powder scattered on the water surface then reveals huge "circumpolar" loops similar to the air waves in the upper westerlies. Recent studies of the atmosphere's circulation have also benefited enormously from the development of computers. Right, a machine used by the American Weather Bureau for the rapid plotting on a map of data based on computer-processed readings.

# 5 Evaporation

The atmosphere expresses many of its processes in visible form, but it needs training to decipher this sky writing. To do so, we must not only understand the grammar of the atmosphere—the general rules that govern its behaviour—but also be able to interpret the meaning of individual messages as they appear in the sky. Several different factors may have combined to produce each atmospheric feature, but an experienced observer can soon separate the major causes from the irrelevant trimmings, and incorporate his findings into a reasonably accurate account of what is happening, and what is likely to happen later.

Clouds are our principal clues, and their form and distribution can tell us a great deal, particularly about the way in which the air has risen to create them. The basic types of cloud have been classified, mostly under the Latin names first suggested by the British pharmacist Luke Howard in 1803. Howard divided clouds into three main types: *cirrus* (a tuft of hair or feathers in Latin), *cumulus* (a heap), and *stratus* (a layer). Cloud classification is based on altitude as well as shape. Cirrus, for example, occurs only at very high levels where temperatures are so low that tiny water droplets are frozen into ice crystals, which are then dispersed by strong winds to give the clouds a fibrous, windswept appearance.

No two clouds are exactly alike and no cloud stays the same for very long. Sometimes the sky is covered with sheet-like stratus cloud that stretches for hundreds of miles. If these clouds are very high—about

Above right, a diagram of the hydrological cycle—the ceaseless exchange of more than 475 million million tons of water a year between the earth and atmosphere. All precipitation (solid lines) is the product of evaporation (dotted lines) from water and land surfaces and of transpiration from plants. Evaporation is particularly intense from lakes in warm, arid regions. Right, a potash extraction works on the shores of the Dead Sea, where evaporation amounts to approximately 10 feet a year, leaving dissolved salts behind.

20,000 to 30,000 feet—they are probably made up of ice crystals; they are then known as *cirrostratus*. Lower stratus clouds, on the other hand, at only a few hundred feet, are composed of tiny water droplets, and light rain may fall from them for several hours. Stratiform clouds usually appear when there is a steady but gradual ascent of air. Where there is patchy overturning in the air combined with rising and sinking air pockets, however, isolated puffs form in the rising, cooling air currents. These are cumuliform, or "heap," clouds.

Different types of cloud often follow one another. For example, a mass of stratiform clouds is often replaced by cumulus clouds a few hours after a travelling depression has passed over. Again, certain parts of the world are dominated by one particular type of cloud: in polar regions, for example, clouds are mainly stratiform, while in tropical and equatorial areas they are

precipitation

condensation

water vapour

evaporation

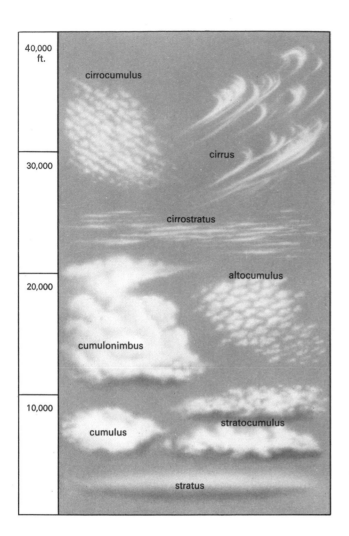

| | |
|---|---|
| 40,000 ft. | cirrocumulus |
| | cirrus |
| 30,000 | cirrostratus |
| | altocumulus |
| 20,000 | cumulonimbus |
| 10,000 | stratocumulus |
| | cumulus |
| | stratus |

Above, a diagram showing the types of cloud we can expect to see at different heights. Alto- means that the cloud is high, and nimbus or nimbo- that it is (or shortly will be) precipitating. The higher clouds are composed of ice crystals and have a windswept appearance; lower clouds are made up of cloud droplets and have a more rounded form.

Right, photographs of the three basic types of cloud: from top to bottom, feather-like streaks of cirrus over south Wales; isolated puffs of cumulus above the Pacific; and low-lying layers of stratus over the Faroe Isles.

Photographs illustrate some of the stages in the development of a cumulonimbus into an anvil-shaped cloud set out in the diagram below. Above the condensation level, a rising column of warmed air forms into a cumulus cloud. Cloud droplets that become too heavy for the up-draught to support begin to fall as rain, dragging cold air down with them. Meanwhile droplets in the upper parts of the cloud have frozen and the cloud has been flattened into an anvil head at the base of the stratosphere.

30,000 ft.

ice-crystal level

20,000 ft.

10,000 ft.

condensation
level

Left, an evaporimeter, which is fixed to a corner of an evaporation tank in the way shown below. It is used to measure the amount of evaporation from a water surface of known area during a specific period of time. As the float inside the open cylinder sinks with the water level, it operates a pointer, which then registers on the dial the number of inches lost.

generally cumuliform. In fact, cumulonimbus—the most spectacular form of cumulus cloud—is particularly common in the tropics. Cumulonimbus is formed when parcels of warm air surge up to heights of 10 miles or more. At this height, their crests freeze into an ice-crystal froth, which blows out in a huge anvil-shaped mass stretching many miles ahead of the main tower. In tropical cyclones, hundreds of such clouds spiral toward the central eye of the storm.

Clouds represent one step in the process by which water is continuously shuttled from the earth's surface to the atmosphere and back again. This cycle—called the *hydrological cycle*—is the most important of all weather sequences. We have already seen in Chapter 4 how water plays a major role in the gigantic task of transferring energy from low to high latitudes. But water is just as important on a more local scale: it controls how wet the soil is, how bright the day, and how often and how intensely it rains. In other words, it provides the core of most people's weather experience.

Water is present in the atmosphere in all three states of matter: gaseous, as invisible water vapour; liquid, as clouds, fog, and rain droplets; and solid, as ice and snow. As vapour, water enters the atmosphere after evaporation from water surfaces, soils, and clouds, and transpiration from plants. (As we saw in Chapter 3, just under half the sun's energy that reaches the earth is used in evaporation—a process that is most intense in tropical areas.) Yet in spite of the enormous importance of evaporation in determining regional and local climate and weather, it is almost impossible to measure it directly; there is no single instrument for recording evaporation rates from seas, lakes, and soils, and transpiration rates from crops, bushes, and trees. Evaporation pans (or tanks) and lysimeters can, however, give some indication of local rates.

Evaporation pans are used to calculate the evaporation from large areas of open water. Water evaporates from a small pan at a uniform rate, but evaporation from a lake or reservoir decreases with the distance from the up-wind edge. The rate of evaporation from large expanses of water is therefore estimated to be about one quarter less than that from a pan at the same temperature and exposed to wind of the same speed. Lysimeters measure evaporation indirectly, by recording the amount of water percolating through an isolated section of soil (with its vegetation). Both these instruments are difficult to use, and their results hard to assess. Evaporation rates are therefore usually calculated from formulas involving components that are more easily measured.

Several different formulas for calculating evaporation from the ground have been worked out, all of which assume one or more of three conditions. First, there must be adequate water to be evaporated and to be transpired—that is, there must be enough moisture to prevent the soil from drying out and so checking the rate of evaporation. To satisfy this condition, a *root constant* is defined: this is the maximum amount of moisture, measured in inches, that can be taken out of a saturated soil without checking transpiration. Amounts depend on the form and age of the vegetation and on the type of soil. The value for grass is 3 to 5 inches of water, but for trees with more deeply penetrating roots it may be 8 to 10 inches or even more. Second, there must be enough energy to provide the latent heat necessary to convert water into vapour. And third, there must be enough air movement to remove saturated air from an evaporating surface, otherwise evaporation and transpiration would come to a halt. This is why a housewife hopes for windy as well as dry weather when she hangs out the washing.

Most physical theories of evaporation base their approach on one or both of the last two conditions. For example, it is possible to draw up an equation in which evaporation is the only unknown quantity based on the principle of energy balance—that there is an equal input and output of energy in the atmosphere. Let us imagine that during summer in the Northern Hemisphere (May to September) 100 units of energy reach the earth from the sun: of these, about 34 are radiated back by the upper atmosphere, 20 are reflected by the earth's surface, 4 heat the air, 2 heat the soil, and 1 is used in photosynthesis. It follows then that about 39 units are used in evaporation and transpiration. These two, called for convenience *evapotranspiration*, therefore absorb more energy than any other single process.

Having calculated the amount of energy needed to convert water into water vapour, it is now possible to estimate the *rate* of evaporation. The only uncertain factor in this break-down is the ratio of energy used in transpiration to that used in heating the air, but this can be found by employing a different method of calculation, which is known as the *aerodynamic* method.

The aerodynamic approach is more complex than the energy-balance method. It is based on equations defining how much water vapour and heat are transported by atmospheric eddies or turbulence away from surface water. As far as water vapour is concerned, amounts depend on the wind speed and the difference in temperature and humidity between the water's surface and air above it. The terms involved are complex and difficult to measure. Nevertheless the British soil physicist H. L. Penman has managed to work out an empirical formula for evaporation from an open water surface.

Rates of evaporation from an open water surface and from a surface covered with vegetation are surprisingly similar at times of intense evapotranspiration. But since transpiration takes place almost entirely during daylight hours, the *total* rate over a 24-hour period is always less than that of evaporation alone from an open water surface. So the longer the hours of daylight, the closer the two figures become. From May to August in the Northern Hemisphere, the ratio between the two is about 0.8, and from November to February, when plant growth is less active, it is about 0.6. These ratios can be used to work out the potential transpiration from a closely planted surface that is the same colour as grass and has an adequate supply of water. The type of vegetation is much less important than might be supposed; the type of soil is also largely irrelevant, unless it affects the root constant.

The aerodynamic and energy-balance methods have been combined by Penman in a formula that expresses the rate of evaporation from an open water surface in terms of elements that can be measured more easily—mean air temperature, mean air humidity, duration of bright sunshine, and mean wind speed. Even so, the calculations involved are lengthy. Less laborious calculations are necessary in the formula, arrived at by the American climatologist C. W. Thornthwaite, for potential evapotranspiration when water supply is ample—that is, when the soil is not dried beyond the root constant.

Recent hydrological experiments show that there is often a rapid fall-off in evaporation in the surface layers of soils after the loss of about the first inch of water—far less than most root constants. Almost certainly, this decrease is due to the shrinking size and number of the channels in which capillary action can take place. So, in strong sunshine, a bare soil acts as its own evaporation regulator. This fact is exploited in dry-farming areas, in which supplies of ground water are built up during years in which the soil is left fallow and kept clear of all vegetation. Keeping the ground clear gives a double benefit: it conserves moisture that would otherwise be transpired by weeds, and it prevents soil fertility from being wasted on unwanted vegetation.

Studies of evaporation have also helped to put irrigation on a sounder footing. In many parts of the world, simple sets of rules have been drawn up that can be applied by farmers to work out their own water budgets. Irrigation is no longer practised only in deserts, and the word aridity has acquired a wider meaning. Arid regions are now defined as those in which the drying of soil below the root constant either impairs growth or causes plants to adapt their physical structure to restrict transpiration losses or even to store water—like, for example, a cactus. According to this definition, large areas that were once thought to be hydrologically balanced or even humid are now classified as arid at certain seasons. And in some of these areas, such as south-eastern England, irrigation has greatly improved the crop yield.

Evaporation studies have also helped to conserve water in reservoirs. One reservoir at Broken Hill, Australia, with a maximum depth of only 16 feet, was losing an average of 7 feet a year before evaporation was checked by spraying the surface with cetyl alcohol, which spreads into a protective monomolecular film. A tremendous amount of water evaporates from open lakes in warm, arid regions. Losses from the Dead Sea, for example, are four to five times greater than the average annual rainfall. Even the more northerly Caspian Sea loses an estimated 5 to 6 inches a year—an amount equal to over a million million gallons. And in so-called temperate climates, there is often a wasteful loss of water from lakes and reservoirs—so much so, that the tradition of combining catchment areas and forestry has been challenged. In wooded areas, too, the water cycle may be further complicated by losses not only from transpiration but from the direct evaporation of rain that is intercepted and caught in leaves and cones. Pine trees can suspend between 20 and 50 per cent of the rainfall in their crowns; deciduous trees retain less, because water runs off their leaves. Evaporation studies have also been useful in areas where trees have been planted to help drain marshlands, as in the Landes area of south-western France.

Evaporation is thus vital as a drying agent and as a link in the hydrological cycle. As a factor of local humidity, it is less significant than is sometimes supposed, except in calm weather. Another popular misconception is that there is more precipitation as well as humidity near quite small areas of water. Water vapour is of course taken up by air moving over water, but in most cases this represents only a tiny fraction of the air's overall vapour content. Even the largest lakes contribute only a relatively small amount to the total amount of moisture involved in the world's hydrological cycle, while millions of tons of water vapour drift invisibly above the world's great deserts. As we shall see in the next chapter, air movements that *condense* the available water vapour are far more important than any local increase in vapour content.

Above, fields of spring wheat in Montana alternate with land left fallow, partly to conserve moisture. Evaporation studies have helped to improve crop yields in both dry areas and those once thought to be humid. Below, sprinklers in a Kent orchard.

Below, water trapped in pine needles. Moisture is lost from wooded areas not only by transpiration but by the direct evaporation of rain caught in leaves and cones. Conifers can intercept from 20 to 50 per cent of the rainfall.

# 6 Condensation

At any moment we care to choose, the atmosphere contains over 13 million million tons of moisture. This is a constant figure since losses from precipitation are always eventually balanced by gains from evaporation. The rate of exchange amounts to the colossal figure of about 16 million tons a second or 505 million million tons a year. Even so, if all the water vapour in the atmosphere suddenly condensed and fell, it would only amount to about one inch of rain over the 200 million square miles of the earth's surface—in other words, the atmosphere contains only the equivalent of one inch of rainfall at any one time. Since the average annual rainfall over the earth is 36.6 inches, it follows that there are roughly 36 evaporation-precipitation cycles each year. This means that a molecule of water vapour remains in the atmosphere for an average of about 10 days.

Once evaporated, a molecule of water vapour usually drifts tens to hundreds of miles before it is condensed and perhaps precipitated back to earth again. Water that falls as rain, snow, or hail on high ground in western Europe has probably travelled about 2000 miles eastward from the North Atlantic. In the interval between its conversion from liquid to vapour over the western Atlantic and its precipitation again on to the land several processes have taken place.

Molecules of moisture are transferred from the warm surface of the Atlantic to parcels of warm, moist air, which later rise through the surrounding cooler (and so denser) and drier air. If there is strong turbulence, or mixing, the properties of both types of air are fairly uniformly distributed up to a height of several hundred feet. But whether or not the air has been mixed, it contains a great deal of water vapour—about 5 per cent by volume. It is also lighter: first, because it has been warmed and forced to expand; second, because a cubic foot of vapour weighs approximately two-fifths less than a cubic foot of dry air at the same temperature and pressure. It follows then that moist air is lighter than dry air, and warm, moist air is lightest of all. This is a fact of considerable importance, as we shall now see.

Above right, a colossal bank of cumulus cloud over the French countryside. Most clouds are produced by the chilling of rising air. As it rises, warm moist air expands and cools; cooling reduces its ability to hold moisture, which then starts to condense into cloud droplets. Right, a diagram showing the place of condensation in the hydrological cycle. Water that evaporates from the sea may travel hundreds of miles in the form of cloud droplets before it eventually falls back to earth as rain, hail, sleet, or snow.

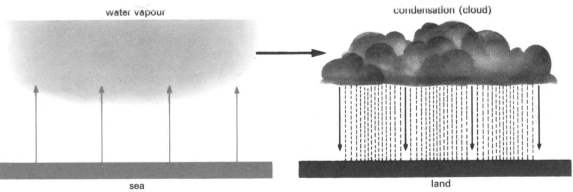

water vapour

condensation (cloud)

sea

land

When air rises, it does so for two reasons: either because it has become lighter as the result of its warmth and humidity; or because it is forced to rise over some obstacle such as a mass of colder, denser air or a line of hills. In either case, it is cooled, since its climb through levels of progressively lower pressure makes it expand. Expansion requires kinetic energy, which is obtained from heat and potential energy in the atmosphere—a process that leads to a drop in temperature. The rate of cooling is, however, often modified by the mixing together of the parcel with the surrounding air.

A quantity of dry air in which there is no condensation or evaporation, and no mixing or other form of heat exchange with the surrounding air, warms or cools at a constant rate as it sinks or rises through the atmosphere. This rate, known as the *dry adiabatic lapse rate* (D.A.L.R.), is approximately 1°c for every 300 feet. But if the air is moist and if condensation occurs within the parcel, latent heat is released, and so the temperature of the saturated air falls much more slowly. The new rate is called the *saturated adiabatic lapse rate*. It is not constant, but varies with the amount of latent heat released—in other words, it depends on the amount of condensation in relation to the temperature of the air. At low levels, where the air is warm and humidity is high, the saturated adiabatic rate is only just over half the dry adiabatic lapse rate. But the rate for saturated air gradually increases with height, and at a very high level it almost equals the dry-air lapse rate. On graphs, D.A.L.R.s are represented by straight parallel lines, while saturated adiabatic lapse rate lines are irregularly spaced convex curves, showing an increasingly steep drop in temperature with height.

The buoyancy of the *moving* air is controlled by its temperature in relation to that of the *surrounding* air. As a general rule, the temperature of the surrounding air decreases with height at an irregular rate—known simply as the *lapse rate*. If a parcel of air is warmer, and therefore less dense than the air around it (and if its vapour content is the same), it rises in the way in which a

Below, warm air rises because it is lighter than the surrounding cool air; also, moist air weighs less than dry air. Conversely, cold, dry air tends to sink.

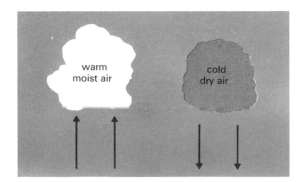

Below, warm air also rises when it is lifted over
some obstacle. This may be a wedge of colder air
or a line of hills. In the above photograph, air
that has been forced to rise over mountains in
Aden has formed a so-called *orographic* cloud.

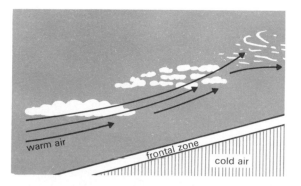

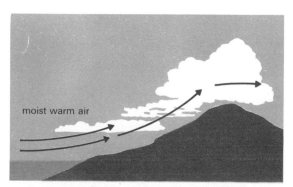

submerged table-tennis ball floats upward through a tank of water. Conversely, when moving air is cooler than the surrounding air, it is denser, and sinks; when it is at the same temperature, densities are equal and the parcel remains stationary or moves only with the surrounding body of air. So atmospheres are divided into those that encourage and those that discourage vertical motions of the air within them.

In ordinary circumstances, if a parcel of air that is moving upward cools to a temperature that is lower than the surrounding air, it naturally sinks back toward its original level. Similarly, air that sinks becomes warmer and rises again. Such an atmosphere is said to be *stable,* since it restricts vertical motion. But if a parcel of air cools or warms adiabatically, so that the rising air becomes progressively warmer and the sinking air progressively cooler than the surrounding air, then its movement, once started, continues at an increasing rate. Such an atmosphere is said to be *unstable,* and encourages vertical motion. If, on the other hand, the

observed lapse rate (usually taken over short distances) is the same as the adiabatic lapse rate for that particular distance, changes in the temperature (and density) of the moving parcel of air are neither encouraged nor discouraged. The atmosphere is then said to be in *neutral stability.*

The lapse rate of a well-stirred atmosphere tends to resemble the dry adiabatic lapse rate, so dry air in the lower troposphere is often in neutral stability. But when there is intense heating near the surface, the lapse rate is probably high, and greater than both the dry and the saturated adiabatic lapse rates. If this is so, the air is unstable with fast-moving currents of rising and sinking air. Thunder-storms, tornadoes, and dust-devils are all manifestations of very unstable air, and these are repeated on a much smaller scale in the sudden whirls of dust above heated pavements in summer. Falls in temperature can be very sharp in the first few inches above ground on a summer's day— many hundreds of times greater in fact than the tropospheric average: a drop of 10°c in

Below, diagrams showing stable and unstable air conditions. On the left, a parcel of air rising and cooling at the dry adiabatic lapse rate (3°C for every 1000 feet) cools more rapidly than the surrounding air in which the temperature falls with height (the lapse rate). It therefore tends to sink back to its original level. Such conditions are described as stable.

On the right, where a parcel of air has been heated, the dry adiabatic lapse rate is less than the environmental lapse rate. The rising parcel of air is therefore warmer than the surrounding air and so continues to rise. When it starts to condense, latent heat is released and the net cooling—the saturated adiabatic lapse rate—is even slower, the air thus rising faster. Such conditions are unstable.

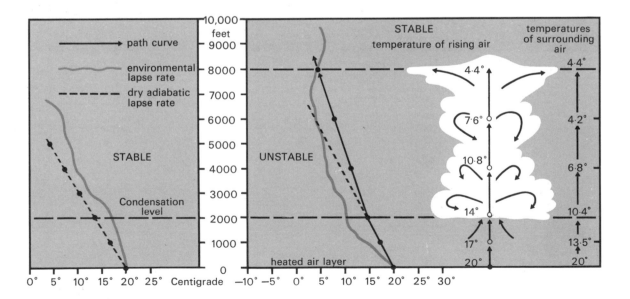

Radiation fog (above) is really a stratus cloud that forms at night near the ground. It occurs when a shallow surface layer of air is chilled to condensation level. This drop in temperature is caused by the transfer of heat from air to the ground, which has already cooled by radiation. In summer, radiation fogs soon evaporate after sunrise, but in winter they may persist all day.

the first 12 inches of air above an English lawn is quite common.

Atmospheres are stable when the observed lapse rate of temperature with height is less than the adiabatic lapse rate. Such conditions are common at night when the earth cools by radiation, and the resulting downward transfer of heat by conduction and radiation from the air to the earth lowers the temperature of the surface atmosphere. When skies are clear and the air is calm, the effect of the surface cooling is often so intense that the air is coolest at ground level and gets progressively warmer with height in the first 500 to 1000 feet. This is known as an *inversion of temperature*; it occurs on about two nights

out of five in western Europe, usually in fine weather. If temperatures drop to the condensation level, the water vapour condenses into droplets, and the resulting cloud that forms in the chilled air near the ground is trapped beneath the warmer air above. This cloud is called a *radiation fog,* but it is only a normal cloud in an abnormal position. Over cities, however, the inversion layer can also trap pollution in the form of smoke and exhaust fumes, which, when combined with fog, builds up to a thick, acrid *smog*—a portmanteau word from smoke and fog. But if the inversion layer is well above the ground —say, some hundreds of feet—then stratiform clouds rather than fog accumulate beneath it. So, too, the tops of cumulus clouds drifting in the trade winds in the tropics are levelled off by the *trade wind inversion* that is formed at 6000 to 10,000 feet by air as it sinks and warms in the tropical belt of high pressure. And higher up still, the stable air of the stratosphere may, like an inversion layer, flatten the crystalline heads of the cumulonimbus clouds—or, for that matter, atomic bomb clouds—that boil up through at least 30,000 feet of the unstable troposphere.

As we have just seen, cloud in the form of fog can occur at ground level when moist air gives up so much heat to the earth that its vapour condenses; generally, however, clouds form when moist air cools because it has risen. The vertical air movements by which it rises can be grouped under four headings, each of which produces a fairly distinctive sort of cloud: first, convective movements leading to cumulus types of cloud; second, wide-spread regular ascents giving layered clouds; third, irregular stirring motions also giving layer clouds; and fourth and last, disturbances in air moving over high ground.

Most cumuliform clouds over land are caused by *thermals*—portions of air that rise from a locally warmed surface. Thermals, which are roughly hemispherical or mushroom-shaped, vary in width from tens to hundreds of yards. As a thermal rises, friction between the moving air and the still air outside causes a sort of tumbling movement that is, as it were, constantly turning the thermal

Most cumulus clouds are formed by volumes of air—known as thermals—rising from locally warmed surfaces. Artificial thermals may be produced by any sufficiently large source of heat, such as the factory cooling towers in the above photograph. Below, a diagram indicating the mushroom shape of a typical thermal and the air movements occurring inside and around it.

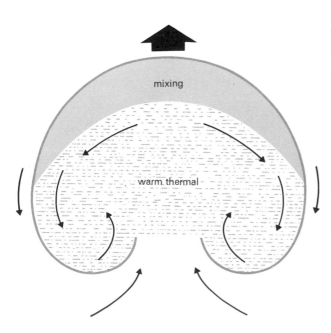

mixing

warm thermal

inside out. Obviously, thermals are invisible below the condensation level, but above it water vapour gradually condenses into a visible cloud of tiny water droplets. As condensation takes place, the release of latent heat warms the thermal and helps to counteract a loss of buoyancy from the mixing of air on its top and sides with surrounding cooler, drier air and also from the evaporation of droplets at the margin of the clouds.

Large cumulus clouds drifting across the sky usually contain several thermals; these rise through the main body of the cloud that is itself partly composed of the remains of earlier thermals. The newer, more active thermals break through the top or flanks of the cloud as growing turrets until, arrested by cooling caused by mixing and evaporation, they either evaporate or sink back into the main cloud tower. The upward movement of thermals may also be halted by a high stable layer such as an inversion, which causes the cloud to spread out horizontally beneath it.

The best view we can get of the ascent of huge layers of the atmosphere is usually near the fronts of mid-latitude depressions. In these areas, there are generally several overlapping layers of clouds, ranging from nimbostratus nearest the ground through stratus and altostratus to cirrus at 30,000 feet or more. (Initially the layers are separated by clear lanes, but these are filled in later.) All these forms of cloud rise very slowly, at four inches a second or even less. This is much slower than the rate of ascent of cumulus clouds, whose rate varies from about three feet a second in the first few hundred feet to 15 feet a second inside small, so-called fair-weather cumulus, reaching as much as 100 feet a second inside large cumulonimbus.

Irregular stirring motions or turbulence above a cool land or sea surface frequently reduce air temperatures to the point where condensation takes place, causing fog or low stratus cloud. On calm, cloudless nights the ground loses a great deal of heat by radiation, so that only a thin layer of atmosphere is chilled and its water vapour condensed. This forms *dew*, the small drops of water that gather on cold surfaces, and especially on leaves, which have a large surface area

Above, a photograph of the cloud formation produced when a layer of moist air is thrown into a series of waves as it passes over a hill. As shown in the diagram below, the crests of the waves that lie above the condensation level condense into near-stationary lee-wave clouds. When the air above them is very dry, these can have a remarkably smooth and bow-shaped appearance.

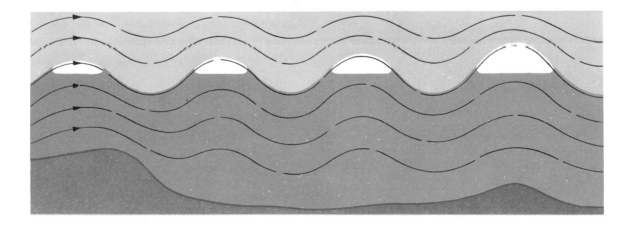

in relation to their mass and so lose heat readily. If the temperature falls below freezing, ice crystals, known as *hoar frost,* are formed. But when there is more air movement, a deeper layer is chilled, and if the air is damp enough or cold enough, a fog forms some 300 to 1000 feet deep. These fogs are, as we saw, called radiation fogs: they occur most often over land areas in autumn and winter, and are most common during anticyclones. This is because anticyclones bring clear skies and light winds, which encourage heat loss; the gentle subsidence of air taking place within them also helps to create near-surface inversions. Fog also forms when warm, moist air flows over a cold sea or land surface. This is known as *advection fog,* and as well as occurring at sea is also common over cold land in winter. When winds are moderate or strong, a deeper layer of air is mixed and the inversion forms at a higher level. If the condensation level is lower than the inversion, a thin stratus cloud nestles between the two. Such low clouds often form in warm airstreams moving poleward in winter, and may bring long periods of drizzle or light rain.

Convection movements may be triggered off in a layer-type cloud when the lower part is warmed by absorbing terrestrial radiation while its top is cooled by night-time radiation. Convection currents are also set up within a layer that is lifted over high ground; as a result, heavy storms may break out over upland areas. Such air movements throw the cloud into a series of waves or bumps, which are visible from below as dark bands or patches. And sometimes the cloud droplets

A series of photographs taken during an experiment simulating the behaviour of a thermal. Heavy fluid containing white powder was released into a tankful of water of uniform density, thus reproducing the rise of a thermal through an atmosphere in which the lapse rate is adiabatic. As the white fluid sinks through the water, it is diluted by mixing and its rate of sinking (a thermal's rising) decreases.

Right, smoke pouring from factory chimneys in a British industrial town. Conversion of vapour into droplets partly depends on the presence of suitable atmospheric particles on which to condense. Large numbers of such condensation nuclei are contained in smoke, and concentrations of several million particles per cubic centimetre are found over industrial areas.

evaporate and clear spaces appear where air is sinking and warming.

When air is forced to climb over a range of mountains, or over a long steep ridge or escarpment, it is thrown into a series of waves at the crest and to the lee of the high ground, rather like the waves downstream of a large boulder in a river. These waves can have an amplitude—that is, height from crest to trough—of 3000 feet or more. Lens-shaped or banded clouds often form in the cool crests of these waves, and when the air above them is stable and dry, their outlines are often strikingly smooth and curved. Hills of less than 1000 feet can disturb the airflow up to heights of more than 30,000 feet, and high mountains can cause wave-like motions as far up as the lower stratosphere—motions that occasionally produce mother-of-pearl clouds of brilliant iridescent colours. These clouds are most often seen in polar regions. A special characteristic of wave clouds is that they remain stationary for several hours, even though a strong wind is continually passing through them. The cloud is, in fact, simply the visible condensation in the rising, cooling limb; the condensed vapour re-evaporates and so disappears in the sinking, warming limb.

Having explained how moist air is lifted and cooled by expansion and mixing, we must now take a closer look at the processes of condensation. As the air cools, its relative humidity increases. Relative humidity is the amount of water vapour that a parcel of air contains, expressed as a percentage of the amount it *could* contain at the same temperature. When air is saturated, its relative humidity is 100 per cent. This state hardly ever occurs in nature without condensation taking place, but it is worth noting that if the air is perfectly pure and clean, it can be overloaded, or *supersaturated,* so that its relative humidity in a careful laboratory experiment can be as much as 700 per cent. In fact, however, the air around us is never free from particles: there are always impurities, called *aerosols,* on which water

vapour can and does condense. These are known as *condensation nuclei,* and because of them, there is little supersaturation in the atmosphere. Other particles known as *hygroscopic nuclei* allow condensation to take place even before the air is saturated.

Condensation nuclei are thus extremely important in the conversion of vapour to droplets. Even so, in spite of a great deal of research, they are still something of an enigma. They are, however, known to vary in size from a few millionths to a few thousandths of a centimetre. On average, a cubic centimetre of air contains from 1 million to 40,000 very small (or *Aitken*) nuclei, about 100 large nuclei, and perhaps only one of the so-called giant nuclei. Actual concentrations, however, vary considerably from place to place: air over the oceans and at high levels contains far fewer nuclei than it does over industrial areas where concentrations of several million particles per cubic centimetre are quite common. Weather conditions also play a crucial part: the currents and turbulent stirrings of unstable air scatter nuclei, and stable atmospheres concentrate them. Also many particles may be brought to the ground by rain and the down-draughts associated with it.

Nuclei originate in several ways: factory chimneys, and domestic, forest, and bush fires throw up large numbers of mainly very small carbon particles. Other nuclei are formed by photochemical action or by the combination of water vapour with trace gases, creating such substances as sulphuric acid, ammonium chloride, and ammonium sulphate. The weathering of the earth's rocks and the salt thrown into the air when sea spray evaporates both produce mainly large and giant nuclei. Mineral particles from the soil only act as condensation nuclei when no soluble nuclei are available; on the other hand, the large numbers of salt nuclei from sea spray are extremely hygroscopic. We have only to leave table salt exposed for a day or two to see how it attracts moisture. On the size and concentration of condensation nuclei depend the size and number of the droplets that form on them; these in turn determine the density of fogs and the type of

precipitation that falls from clouds. Clouds that form over the oceans usually contain less than a hundred large water droplets per cubic centimetre, most of them condensed around sea-salt particles. Clouds and fogs formed over the land contain on average a few hundred small droplets per cubic centimetre. Of these only about one tenth have sea-salt nuclei; of the rest, most have nuclei of smoke particles and a few of soil particles.

How fast cloud droplets grow upon their nuclei depends on a number of factors. Most important of these are the size, composition, and concentration of the nuclei; the rate of cooling of the air; and the type of motions in the cloud. Most cloud droplets reach a radius of between 0.005 and 0.01 millimetres. Only a small proportion have a radius of more than 0.02 millimetres, and very few are likely to be larger than 0.03 millimetres. Even a water droplet growing on a giant hygroscopic sea-salt particle would take several hours to reach a radius of 0.1 millimetres. Yet in cumulus clouds a droplet has less than an hour in which to grow, and in layer clouds, where up-draughts are weak, the droplet would fall out before growing to this size. In practice, condensation on the original nucleus seems to come almost to a halt when a radius of about 0.05 millimetres is reached. This is why so few clouds give rain. At the same time it raises one of the central questions of cloud physics: how do tiny cloud droplets combine to form raindrops, snow flakes, and hailstones? Raindrops are about one million times larger in volume than cloud droplets; and, as we have just said, they also fall within an hour of the cloud's first appearance and often after only a few minutes.

We can answer this question only by bringing into the discussion another important change of state—that of liquid into solid. In particular we must see how liquid cloud droplets change into solid ice crystals. But since water droplets freeze even less readily than they condense from vapour, the freezing process is often delayed until temperatures fall well below 0°c. In other words, there is considerable *supercooling.* Clouds with temperatures as low as –20°c are often mainly composed of supercooled droplets. Below about –40°c (called the *Schaefer point* after the modern American meteorologist who has greatly enlarged our knowledge of atmospheric ice) water freezes spontaneously. But at temperatures above –40°c, the freezing of water into ice, like the condensation of vapour into water, depends upon the presence of suitable nuclei. Freezing involves not only an accumulation of water molecules, but also their arrangement in a neat pattern, so the only microscopic particles in the upper troposphere able to act as nuclei for the growth of ice crystals are probably the few that have a crystal structure similar to ice.

So far there is little agreement among meteorologists on the composition and origin of nuclei (other than ice itself) that bring about freezing at temperatures above –40°c. There are two main schools of thought: one school says that freezing nuclei are the products of terrestrial weathering and erosion, and are carried into the upper troposphere by convection currents; the other explanation is often associated with E. G. Bowen, an Australian meteorologist who believes that freezing nuclei originate outside the earth and lodge in the troposphere after meteoric storms. Bowen supports his idea by claiming a relationship between meteoric showers and increased rainfall in Sydney 30 days later. The interval between the two events represents the time taken for the meteoric dust to fall into the troposphere from heights of 50 or 60 miles. However, Bowen's theory has been severely criticized, and from the evidence we have at present it seems that the nuclei do in fact come from the earth. Clay—in particular kaolinite—is thought to be a major source.

Once freezing starts, a cloud changes quite rapidly into a mass of snow crystals, each of which is formed from several thousand cloud droplets. (When melted, each crystal is the same size as a small raindrop.) We have already seen, from observation, that freezing nuclei are rare and that supercooled droplets tend to freeze on ice—two factors that should in theory produce only a small number of individually large crystals. It is therefore difficult to explain the rapid transformation of

Evaporating sea spray provides a source of large condensation nuclei in the form of salt particles.

supercooled droplets into ice crystals, and the high concentrations of melted snowflakes (formed by an accumulation of ice crystals) that are necessary to produce steady rain from widespread layer clouds. There are two probable explanations of how this can happen in spite of an apparent lack of freezing nuclei. One is that the delicate branches of the familiar star-shaped snow crystal that forms at temperatures between –16 and –12°c are broken off by collisions and air resistance; they then act as nuclei for other crystals. The other explanation is that additional ice nuclei are also created when supercooled droplets freeze. First, a thin shell of ice is formed around the droplet; a fraction of a second later this is shattered as the interior freezes and expands. Water ejected through these cracks then freezes into about 10 ice splinters, which in their turn break off to act as nuclei. In these circumstances, it is not so surprising that the droplets of a cloud should freeze so rapidly.

No two ice crystals are exactly the same. Their shape depends on temperature and on

Top, a branching frost crystal of the type that develops on trees in calm weather; below it is the corresponding snow crystal that forms when temperatures are between –12 and –16°C. Next, a plate-like frost crystal with large fog particles attached to it, and (bottom) the corresponding snow crystals with cloud particles attached. This type of crystal forms at lower temperatures, about –16 to –25°C.

Cirrus clouds (right) are composed entirely of ice crystals and are found at heights of about 25,000 to 30,000 feet. The clouds spread with the prevailing winds; wisps hanging from the clouds are falling drops of rain that evaporate before reaching the ground.

the degree of supersaturation of air with respect to ice, which increases as temperatures fall. At relatively high temperatures, large plate-like crystals are created by slow growth on a few active nuclei; these group together to form large snowflakes. At lower temperatures, many other nuclei are active and these then produce large numbers of branching or needle-like crystals. Such crystals do not easily melt since temperatures beneath the cloud are still low, and so they fall as fine, powdery, drifting snow.

An interesting feature of the ice-water-vapour triangle is that moist air is saturated at one temperature in respect of ice and at another in respect of water. For example, air is 100 per cent saturated in respect of ice at $-20°$C, while at the same temperature it is only 84 per cent saturated in respect of water. One result of this is that ice crystals can continue to grow in air that is not humid enough to produce water droplets. Another is that an ice crystal, once grown, can fall through air that is only 68 per cent saturated without losing any of its bulk by evaporation. Also,

since ice does not easily evaporate, the margins of ice clouds are generally vague and blurred, in contrast to the edges of droplet clouds, which are nearly always clear-cut because of the rapid evaporation of any droplets that wander into the unsaturated air outside. A switch from a clear to a vague outline is, in fact, the best way of spotting that freezing is taking place in supercooled-droplet clouds. When this transformation—known as *glaciation*—occurs, the tops of large cumulus clouds change into an anvil-shaped mass of drifting and falling crystals. For the reason we have just mentioned, anvils survive for many hours after the main body of the cloud has evaporated or fallen as rain. These high, crystalline cloud remnants are called *anvil cirrus*.

Such then are the mechanics of condensation of water vapour in the atmosphere—sometimes into water droplets, and at others into ice crystals. We now go on to see how the condensed particles eventually descend upon the earth as rain, hail, or snow in the process we call precipitation.

# 7 Precipitation

Precipitation occurs when water droplets or ice crystals grow so heavy that the speed at which they fall exceeds the speed of the upward-moving air. After crystals leave the cloud, they continue to grow for a short time in the moist air. But soon they, like raindrops, begin to evaporate. Some of the drops are large enough to reach the ground as precipitation, but the rest form trailing wisps, or *virga,* that hang like a ragged curtain beneath the cloud. Whether raindrops survive without evaporating depends on their size and on the relative humidity of the subcloud layer.

The British meteorologist B. J. Mason has calculated that a droplet with a radius of 0.01 millimetres (a cloud droplet) would fall only 3 centimetres (1.25 inches) before evaporating in air of 90 per cent relative humidity. A droplet with a radius of 0.1 millimetres would fall 150 metres (480 feet) and a 1-millimetre-radius droplet about 40 kilometres (24 miles). Since most cloud bases are a few hundred metres above ground, a radius of 0.1 millimetres is usually taken as the minimum size at which a droplet can reach the ground as precipitation. On this basis, drizzle is defined as a mass of droplets with a radius between 0.1 and 0.25 millimetres. A raindrop cannot grow to a radius of much more than 2.5 millimetres without becoming deformed and broken as it falls. A typical raindrop has a radius of about 1 millimetre and falls with a terminal speed—that is, the speed at which the acceleration due to gravity is balanced by air friction—of 6.5 metres per second; snowflakes, with a much larger surface area, flutter down at about 1 metre per second.

Drizzle generally falls from low, shallow layer clouds with a ragged base, while larger drops, with radii of up to 2 millimetres, fall from deeper layers, such as those associated with warm fronts. The heaviest rain falls from cumulus clouds several miles deep, though usually in the form of sporadic showers, since the active life of an individual precipitating cloud is less than one hour. When layer clouds extend well above the 0°C level, ice crystals form and combine as snowflakes, which in cold weather reach the ground as snow. In warm weather, they melt and fall as rain. Often, vigorous currents in cumulus clouds amalgamate the ice particles into hail pellets, which may reach the ground either as *graupel* (soft hail) or as rain.

Above right, a vast cumulonimbus cloud sheds some of its load of water as rain. All clouds are made of condensed water vapour, but only some are capable of precipitating. In this chapter, we review the processes by which cloud droplets grow into the larger rain droplets or ice crystals that eventually fall as rain, hail, or snow.

Right, a world map showing average annual rainfall measured in inches.

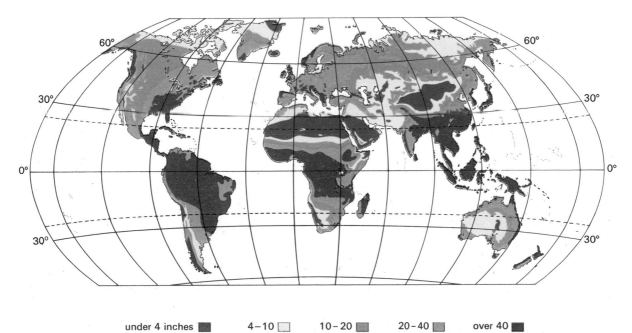

under 4 inches ■  4 – 10 ☐  10 – 20 ■  20 – 40 ■  over 40 ■

85

Top, rime that has formed on the windward side of branches. Above, dew is water vapour in the air that has condensed on contact with a surface cooler than the surrounding air, such as here on lambs' wool caught on barbed wire.

Snowflakes are loose clusters of up to 100 ice crystals. These crystals are united by the fairly gentle movement of air in deep layer clouds that extend well above the freezing level in winter. The flakes usually melt rapidly after they fall below the freezing level, but if the air is dry, evaporation keeps them cool and frozen even when the near-surface temperature is as high as 3°C. When temperatures are between 3 and 4°C, sleet falls; but some snowflakes may still reach the ground when temperatures are between 4 and

Left, photographs of an artificially produced snow crystal. Top, a hexagonal plate formed in air of low humidity and low temperature ( 19°C). The next two pictures show the development of typical branching extensions when the humidity was abruptly increased, indicating a fairly sharp transition from dry to humid air when this type of snow is falling.

Above, a hailstorm in Texas. Hailstorms develop most often over the interior of continents during summer. Thermals rise from the warmed earth because the air is unstable; above the condensation level, huge cumulonimbus clouds build up with strong upward currents—conditions that favour the formation and release of hail.

7°c if the air is very dry. Melting snowflakes and evaporating droplets need latent heat to change their state, and this they absorb from the air through which they fall, so that the air is cooled; cold air is also dragged down with them. As a result, temperatures may fall enough for the snowflakes that follow rain and sleet to reach the ground without melting.

In cumulonimbus clouds, where there are vigorous currents and a high water content, supercooled droplets at temperatures below about –20°c may collide together to form

the loose masses of tiny frozen droplets that are known as graupel, soft hail, or snow pellets. These occur mainly in winter, when in low temperatures the tiny supercooled droplets freeze together on contact into a growing mass of ice with a great deal of air trapped between the frozen droplets. In a similar way, *rime*—which is the accumulation of loosely packed white ice—forms on the windward side of obstacles such as tree branches, telegraph poles, and electricity pylons that lie in the path of a supercooled fog.

The origin of large hailstones has for a long time been a controversial subject among meteorologists. This is not surprising, since it is hard to explain how lumps of ice that hurtle downward at speeds of 60 or more miles per hour take only 20 minutes to grow to the size of tennis balls. Part of the difficulty is due to our limited experience of what takes place inside hailstorms—understandably, when one considers the danger of inspecting a hailstorm too closely. Today, however, we can explore clouds with the aid of radar; we can even measure the size of falling hailstones while they are still in the cloud. Hailstones are most common during summer over the interiors of continents in the mid-latitudes, where cloud densities are often high and strong thermals rise from the warm earth—two essential conditions for hail formation, as we shall shortly see.

Hailstones are not always spherical. Many of the largest stones have a disc-like or conical form, and sometimes stones are kidney-shaped with projecting lumps like rounded reversed icicles on top. The largest stones generally weigh from 1 to $1\frac{1}{2}$ lb. The internal make-up of hailstones has aroused a good deal of interest: many are constructed like an onion, with alternating layers of clear and white ice enclosing a nucleus that is often a frozen raindrop. This structure seems to evolve in the following way: first, in temperatures only slightly less than freezing, the growing hailstone collides at high speed with relatively large supercooled droplets and snow crystals. These spread over the surface of the stone before freezing into a skin of high-density, clear ice. Then, under certain conditions, the freezing of the supercooled droplets on the hailstone may release enough latent heat (of fusion) to raise the surface temperature of the stone to 0°C. The stone then becomes covered with a slushy mixture of water and ice, some of which is swept off the upper surface of the stone; the rest refreezes into a layer of dense, clear ice. At temperatures below –20°C, small spherical cloud droplets freeze rapidly after hitting the stone, and since a great deal of air is trapped among them, they form a layer of low-density, opaque white ice.

When hailstones are composed of alternating layers of clear and white ice (and not all are), it follows that the storm-cloud contains air currents vigorous enough to toss the heavy and fast-moving stones back and forth between freezing level and ice-crystal level (that is, –40°C) and so build up a stone within about 20 minutes. Hailstones do not grow, of course, in the crystalline crest of a cloud; these crests also generally spread downward as glaciation proceeds, so that clouds lose much of their ability to form hailstones. At the same time, they are more likely to produce snow and rain. In fact, a single storm moving across country often produces parallel swaths of hailstones and rain.

Hailstorms are often multiple: we experience a sequence of storms that follow one another down-wind. The explanation is that a new storm is triggered off by the scooping action of the nose of cold air that descends through the cloud of the previous storm and then fans out ahead. This cold, sinking air is responsible for the fall in temperature, gusty winds, and rise of pressure that herald each storm. If a number of storms occur simultaneously, the sinking air may even form a small anticyclone.

Once it had been proved conclusively that condensation alone could not produce precipitation particles, many alternative rain-making processes were suggested. Most of these, though possible in theory, were found to be too slow to work in practice. Precipitation forms in 20 minutes or less in all clouds except a simple layer cloud (1 hour) or a multi-layered cloud (20 hours, as in a depression). Yet several of the rain-making processes proposed for cumulus clouds would take several days. In 1933, however, the Norwegian meteorologist Tor Bergeron arrived at one explanation of rain formation that is still generally accepted. This process has been called after him, though it sometimes also bears the name of Walter Findeisen, the German meteorologist who further developed Bergeron's idea in 1938.

Bergeron based his theory on the interaction of supercooled droplets and ice crystals in the top of clouds below 0°C. Since air in such a cloud is undersaturated with respect

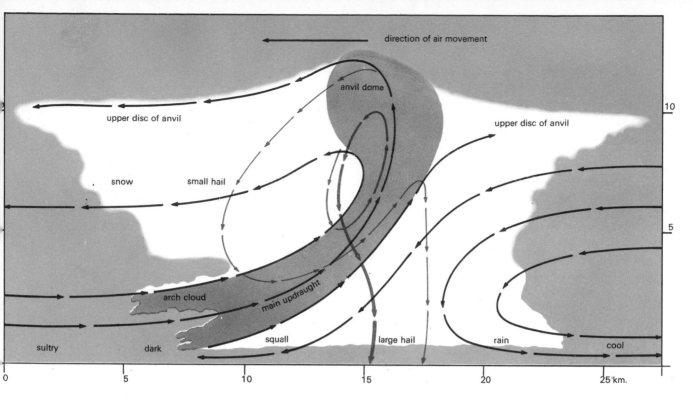

direction of air movement

anvil dome

upper disc of anvil

upper disc of anvil

10

snow

small hail

5

arch cloud

main updraught

sultry

dark

squall

large hail

rain

cool

0       5       10       15       20       25 'km.

The accompanying diagram shows the patterns of airflow in a hailstorm as proposed by F. H. Ludlam, and it accounts for all the observed features. The largest hailstones are thrown out of the core of the strong up-draught and grow as they fall before re-entering the up-draught at a lower level. Some will be too large and have too fast a fall speed to re-enter and, lifted again and again, these fall out as medium-sized hail. The rest will be lifted for a second time, eventually growing large enough to overcome the up-draught just below the crystalline anvil and to fall almost vertically as large hail, gathering speed to hit the ground at up to 100 miles per hour.

Below left, a section through a large hailstone, showing its structure of alternating layers of clear and white ice. Below, a section through another giant hailstone photographed with polarized light, which clearly illustrates the variously sized crystals in the different layers.

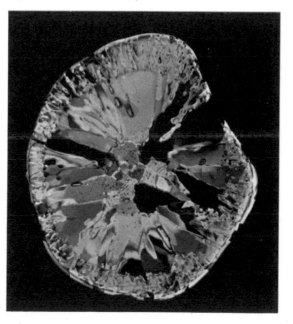

to the water droplets and supersaturated with respect to the ice, the droplets evaporate and condense on the ice crystals. These grow in the strong up-draughts to a diameter of several millimetres in 10 to 30 minutes. They then fall through the cloud until they melt to produce large raindrops. As we have already pointed out, rain often follows the change of cumulus into cumulonimbus with a crystalline anvil. The process is most likely to occur at cloud temperatures from $-10$ to $-30°$c. At higher temperatures, there are not enough active freezing nuclei, and in colder clouds there are so many active nuclei that individual crystals and droplets are too small to reach the ground. Finally, if the water content of the cloud is too small, other rain-forming processes come into play.

So the Bergeron-Findeisen process, though important in explaining heavy, showery precipitation, is not the only way in which rain is made. Often rain falls from droplet clouds and sometimes from clouds whose entire mass is at temperatures above $0°$c. This means that in certain conditions rain must form without an intermediate freezing process. To explain this, the British meteorologist F. H. Ludlam and the Australian E. G. Bowen have both analysed the formation of showers in terms of the so-called droplet-coalescence process: equal-sized droplets, such as those that occur in fogs, do not coalesce easily, but a large droplet with a radius greater than 20 microns coalesces with or sweeps up smaller cloud droplets in its path. As a droplet grows, its coalescing power increases, particularly toward the end of its growth period. Most clouds contain droplets that vary in size because of differences in their evolution and in the nature of their condensation nuclei. Since large droplets have higher natural fall speeds than small ones, they are constantly colliding and coalescing to form droplets of raindrop size in from 20 minutes to 1 hour. Clouds that have developed over the ocean provide particularly favourable conditions for this process because of their large, hygroscopic salt nuclei; so do those with a large water content such as the deep, persistent cumuliform clouds that occur in summer.

Moderate up-draughts of 1 to 5 metres per second also play a helpful part; even weak up-draughts in summer cumulus or stratocumulus clouds are capable of producing light rain by coalescence. In very shallow (3000 feet or less) stratus and stratocumulus clouds, on the other hand, the combination of very weak up-draughts and small water content yields little more than very light rain or drizzle by coalescence. This precipitation may, however, be increased by cooling as the cloud rises over highland areas. Much of the precipitation in North America and Britain falls from multi-layered clouds in just this way. Ice crystals that have fallen from the upper part of the cloud melt, and then grow in the lower parts by coalescence—a combination of the Bergeron and coalescence processes that may result in moderate to heavy amounts of rain.

With this description of the processes of precipitation, we complete our study of the hydrological cycle—the continuous exchange of water between the earth and atmosphere. Generally, precipitation takes a normal course; but there are other phenomena associated with it that are the exception rather than the rule, and many of these are noisy, spectacular, and dangerous. In the following chapter, we shall be giving examples of some of these extreme conditions.

An adaptation of a diagram worked out by B. J. Mason. It outlines the processes by which cloud droplets and ice crystals in layer clouds (left) and cumulus clouds (right) develop into the various forms of precipitation, which are marked with their appropriate meteorological symbols.

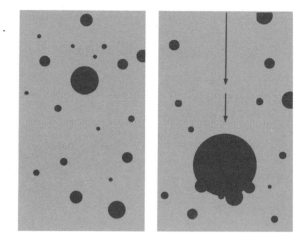
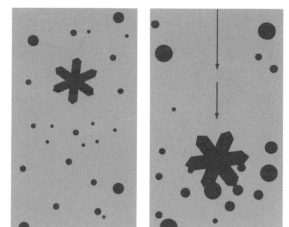

Above, two processes by which a cloud droplet and a snow crystal grow into rain or snow. In the first picture, droplets of various sizes have formed by condensing on nuclei. Second, a large droplet grows by collision with small droplets until it is heavy enough to fall as rain. Third, a snow crystal forms on a freezing nucleus while surrounding droplets evaporate and so provide vapour for the crystal to grow. Fourth, the crystal continues to grow by coalescence with the droplets and either melts into a raindrop or reaches the ground as snow.

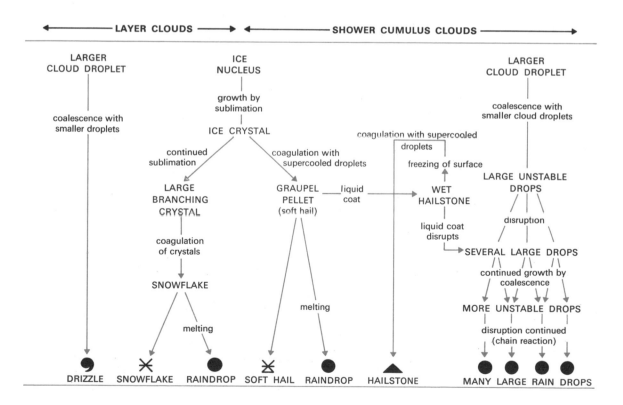

# 8 Light, Sound, and Fury

Lightning is a gigantic spark between a cloud and the ground (most likely when the cloud base is low), between two separate clouds, or even between different areas of the same cloud. It occurs when the build-up of static electricity exceeds the air's natural resistance to the passage of a current. Resistance is high in dry air, but when the air contains water droplets, a voltage of 10 million volts will give a flash. In England, about two thirds of all flashes are single-cloud flashes; in South Africa, the proportion is nine tenths. And despite their appearance, there is in fact no difference between "forked" and "sheet" lightning. Sheet lightning is merely a diffuse glow seen when the usual forked flash is obscured by rain or cloud. In clear conditions, lightning may be visible for a distance of 100 miles.

Lightning discharges a great deal of energy —approximately $10^{10}$ joules—in less than a thousandth of a second. Three quarters of this energy is expended in heating the air that conducts the current to a temperature of about 15,000°C. It is this incandescent air, not the electricity that heats it, which we call the lightning flash. Such sudden and intense heating, by causing the air to expand very sharply, produces a shock wave, the sound of which is *thunder*. A lightning discharge directly overhead causes a single ear-splitting bang. But from a distant flash, the same noise makes a rumbling sound that lasts several seconds. This is because we hear the original thunder-clap as it is reflected back to us over various distances—from clouds, from the ground, or from buildings in a city.

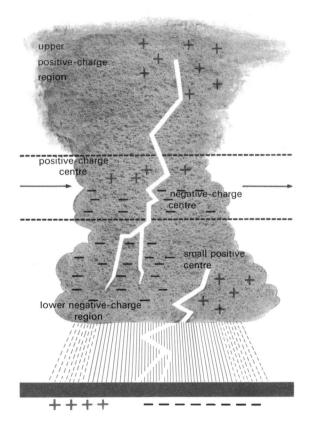

Above, a diagram showing the distribution of electrical charges inside a typical thunder-cloud. The positive crest and negative centre layer are probably created when supercooled droplets freeze; positively charged ice splinters are then carried to the upper part of the cloud and negatively charged frozen nuclei to the lower. The positive and negative areas in the cloud base perhaps result from the disruption of raindrops into positive drops and negative spray. Right, lightning is the visible result of electricity discharges between the cloud and the ground.

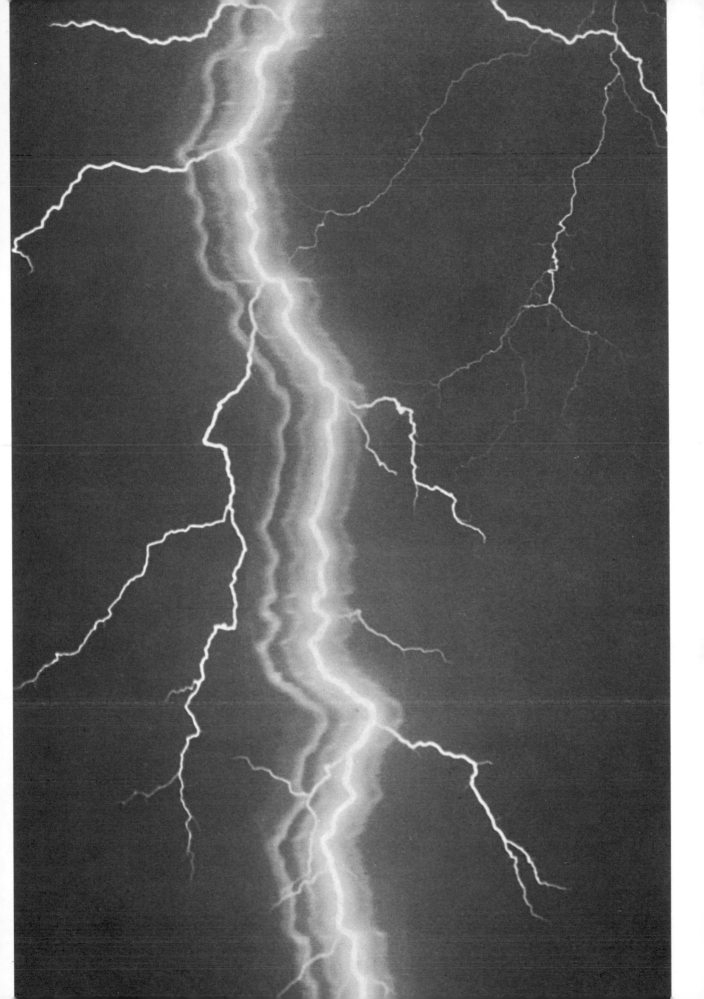

Observations show that thunder-storms usually occur in association with cumulonimbus clouds with crystalline heads and strong up-draughts. The top parts of these clouds are positively charged with electricity; the middle and lower parts are negatively charged, except for a small positive layer near the base. This charge distribution in the lower part of the clouds induces a mainly positive charge on the ground immediately under the cloud, though it is negative away from the storm and in a small area beneath the isolated positive charge in the cloud base.

The first flash of lightning generally occurs from 10 to 20 minutes after raindrops first appear as echoes on the radar screen. The origin of the charge and its distribution have been continuously debated ever since the American Benjamin Franklin—at great personal risk—first showed in the 1730s that lightning is in fact electricity. The small lower positive charge may originate in the disruption of large raindrops, with the result that large drops carry the positive charge down, and the fine spray from the surface of

the original droplet takes the negative charge up. Even so, this theory still does not explain the much larger pattern of positive crests and negative bases in thunder-clouds. This particular pattern seems to be closely related to the glaciation of cloud tops. An important step in the process probably takes place when supercooled droplets freeze; the frozen nuclei then take a negative charge, and minute splinters take a positive charge. (As described in Chapter 6, the splinters are broken off and carried to higher levels.) In such conditions a cloud is capable of producing flashes every 20 seconds or so—the same frequency as that occurring at the height of the storm. Even so, it must be admitted that we still do not entirely understand the electrical and other processes that take place inside thunder-clouds.

Most people, if asked, would probably say that the main, brilliant flash of lightning moves from a cloud to the ground; in fact, it travels the other way round. High-speed photography shows that several down and up strokes follow the discharge path made

Below left, a lightning stroke as it appears to an observer. Right, diagram of high-speed camera record shows what actually happens. A leader stroke from cloud to ground is followed immediately by a return stroke to cloud. This action is repeated several times. The whole process takes less than one tenth of a second and appears as a single flash.

Right, lightning strikes the Empire State Building in New York. Lightning, seeking the shortest path between the cloud and the ground, is most likely to strike the tallest available conductor. Research has shown that the skyscraper acts as a lightning conductor for surrounding buildings; it has been struck as often as 48 times in one summer.

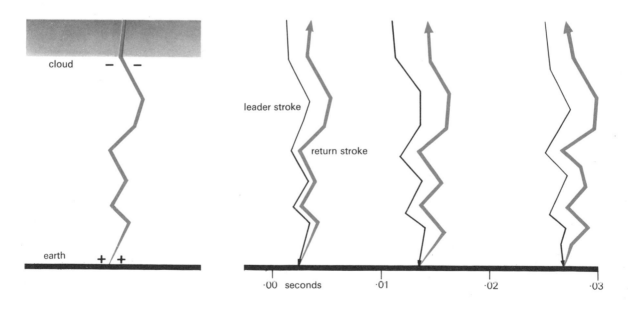

cloud

leader stroke

return stroke

earth

·00 seconds     ·01     ·02     ·03

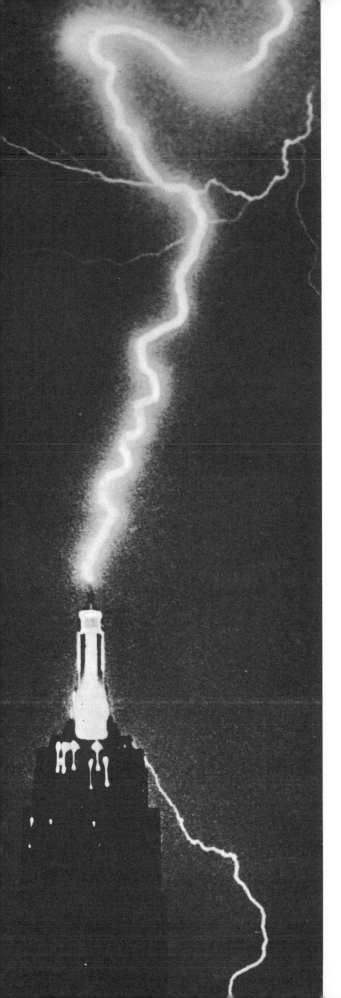

by the first relatively weak leader stroke from the cloud to the ground. Dart and return strokes, as they are called, may be repeated several times to tap different parts of the negatively charged layer in the lower part of the thunder-cloud.

The British climatologist C. E. P. Brooks has estimated that at any one time there are on average 1800 thunder-storms taking place over the earth. All these between them discharge 100 flashes per second, of which about 10 reach the ground. The net result is equal to a constant upward current of about 200 ampères; more powerful still is the current of approximately 1000 ampères produced by a "point discharge" from trees, buildings, and lightning conductors, and consisting of rising streams of positive ions produced beneath the storm-cloud. This upward current in the neighbourhood of thunder-storms is balanced by downward ionization currents in areas of fair weather.

The term *thunderbolt* derives from the devastating effect of lightning, which seemed to behave like something hurled from the sky. Its intense heat can set fire to trees and buildings; it can even shatter them to fragments by the build-up of internal pressure created by its instant conversion of moisture into steam. *Ball lightning* is a glowing ball of air, about one foot in diameter, that drifts above the ground and is said to be liable to explode if it bumps into a solid object. There are few authentic reports of this phenomenon, and its cause is unknown. *St. Elmo's fire* is a luminous point discharge from masts, lightning conductors, and aeroplane wings, where the electric potential is concentrated on a small surface area.

*Tornadoes* contain the earth's fastest winds—probably between 300 and 500 miles an hour. A tornado is a long narrow vortex or whirlwind, only 200 yards or so wide, that hangs below a thunder-cloud, and often reaches the ground. Tornadoes move across country at 10 to 30 miles per hour, but in prairie states of America speeds of 20 to 50 miles per hour are more common. Tornadoes can cause enormous damage, especially in built-up areas. The strong winds themselves are only partly responsible; most of

95

Left, a double rainbow. We see rainbows only when light from the sun behind shines directly on falling raindrops ahead. The sunlight is then reflected and refracted by the drops into the colours of the visible light spectrum. The second bow is a paler, re-reflected image of the first.

Right, a halo around the sun. Haloes are usually seen when the sun is looked at through a cirrus cloud composed of sparsely distributed ice crystals.

the destruction results from the "explosion" of buildings that are crossed by the zone of extremely low pressure at the centre of the tornado. This drop in pressure also lowers the temperature, causing condensation into a funnel-shaped cloud. Powerful up-draughts in the funnel also pick up loose objects and deposit them many miles away; even cars can be lifted and overturned. Fortunately, individual tornadoes are short-lived, but generally they move in groups, like the thunder-storm cells that cause them. Such a group of tornadoes did enormous damage and killed more than 200 people in the American Midwest in April 1965. In Britain,

where tornadoes are comparatively rare, they average only 12 a year, most of which move from the south-west over southern and south-eastern England. In America, where tornadoes are widely known as *twisters,* there are about 150 a year; they are particularly common during afternoons in late winter and spring in the southern states, and in early summer in the Midwest. The risks are so great that tornado shelters are quite common, and a warning system has been organized, which plots the occurrence and movement of tornado packs.

A *waterspout* is a form of maritime tornado that sucks up water from warm seas into the

base of cumulonimbus clouds. Desert *dust-devils* and small whirls of rising dust above heated roads are also similar in form to tornadoes, though they are caused in a different way—by extreme instability in unsaturated air near the ground.

We still do not completely understand how tornadoes originate, but they probably start as small rotations in the air as it rises into the base of growing cumulonimbus clouds. The initial rotation and uplift probably takes place on the fringe of a cold down-draught that sinks to the ground and spreads out from a cumulonimbus storm.

When sunlight or moonlight passes through the ice crystals of cirrus or cirrostratus clouds and is refracted by them, the sun or moon appears to be ringed by a *halo* that varies in colour from red on the inside to white on the outside. Cirrus clouds and haloes often herald an approaching storm.

We see *rainbows* when, standing with our backs to the sun, we observe sunlight that has been both refracted and internally reflected inside raindrops that are falling ahead of us. The light then re-emerges at an acute angle, but split up, as by the prism in a spectroscope, into the component colours of the visible light spectrum. (From ground level, we see only a part of the total reflection, but when viewed from an aeroplane, each rainbow appears as a full circle.) Often there are two rainbows: the primary bow is a series of brightly coloured bands, first violet on the inside, then blue, green, yellow, orange, and red; the colours of the second arc, which lies outside the primary, are in the reverse order because they have been reflected yet again on the raindrops. This second reflection also reduces the intensity of the colours.

Local weather—the passage of clouds and rain, hailstorms and tornadoes—may appear to the unpractised eye as a completely haphazard sequence of events. But synoptic observations and regional mapping carried out since the mid-19th century have shown that there are logical weather patterns on a regional scale that are associated with moving pressure systems; we shall be dealing with these systems in the following chapter.

Far left, a tornado passing over a group of buildings near Dallas, Texas, in 1957. Left, twin water-spouts in a squall in the Gulf of Mexico.

Violent winds and a sharp drop in temperature in a tornado can shatter buildings and scatter debris for miles. Right, wreckage after a tornado at Murphysboro, Illinois.

# 9 Synoptic Meteorology

Local weather is the product of on-the-spot conditions as well as more remote influences. Temperatures, for example, depend both on local heat exchanges and on the warmth imported from elsewhere by the prevailing winds. So to understand changes in weather, we must take stock of conditions over a large area at frequent intervals. For this analysis, we use standard symbols and plot the weather on *synoptic charts*.

At first sight, a synoptic chart may appear enormously complicated. But the jumble of lines, symbols, and numbers represents a series of changing systems that have been classified by meteorologists into idealized weather types or models. These can build up only a general picture of actual weather conditions; even so, they provide a useful key for analysing the weather.

One of the earliest methods of weather analysis and forecasting was based on the idea of *air masses*. These were originally conceived as near-uniform regions of the atmosphere, more especially regions of near-uniform temperature and humidity. Following the work on the subject by Norwegian scientists during World War I, meteorology in the 1930s was particularly concerned with air masses. More recently, however, regular measurements of the upper atmosphere have discredited the idea of complete air-mass uniformity and reduced their importance as a means of interpreting the weather. Climate, on the other hand, is portrayed more and more in terms of air masses, since here the inevitable variations in temperature and humidity are less important.

Today the same system of classifying air-mass types is used (with minor modifications for local conditions) by climatologists all over the world. Air masses are labelled with letters to show their place of origin, whether their source region is maritime or continental, and usually whether they are warm or cold: A stands for arctic; P for polar; T for tropical; E for equatorial; S for superior air—that is, air descending from above, usually in an anticyclone; m for maritime; c for continental; w for warm; and k for cold. For example, the symbol cPk would be used to describe a cold air mass with its source in the high-latitude interior of a continent.

Large-scale air masses acquire their properties from a prolonged stay—often amounting to stagnation—over a source region with distinctive and reasonably uniform characteristics, which are then imprinted upon the

Facts about weather conditions in many areas are gathered at the Central Forecasting Office, at Bracknell in Berkshire. Top left, the office receives teleprinter messages from weather stations throughout Europe. Top right, pictures of cloud over Europe and the Atlantic are received direct from a satellite. Centre left, information from weather stations is plotted on a synoptic map; isobars are then drawn in by hand. The resulting charts are then studied by forecasters (centre right), who analyse the weather trends and issue a forecast. Bottom left, from its Facsimile Room, Bracknell sends out its own charts, and also those received from other centres. Bottom right, a computer makes possible 24- and 48-hour forecasts.

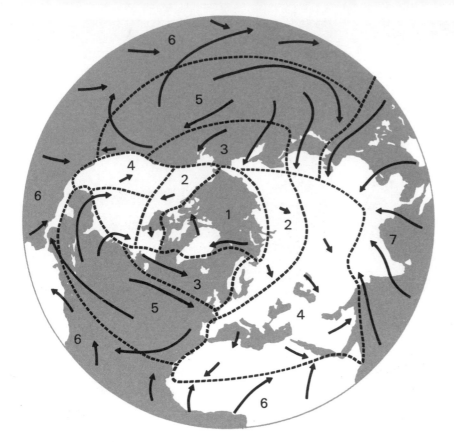

Above, summer air-mass sources. (1) Arctic: this brings
cold, rainy weather to Britain; (2) polar continental:
this is usually warm and dry; (3) polar maritime:
gives showers with bright intervals; (4) tropical
continental: this is comparatively rare in Britain,
bringing hot, dry weather with a characteristic
haze; (5) tropical maritime: conditions are
uncomfortably hot and humid, with sea fog around the
coasts. The other two—(6) equatorial and (7) monsoon
—do not influence Britain's weather.

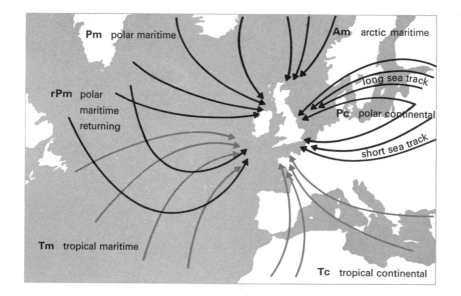

The map on the left shows in more
detail the usual routes by which
the various air masses reach the
British Isles.

Right, cumulus clouds formed
during polar maritime weather;
they build up during the morning,
and by late afternoon the sky is
heavy with cloud, and rain falls.
Far right, tropical maritime
weather often brings sea fog, as
seen here over the Antrim coast,
Northern Ireland. The fog drifts
inland, and disperses.

air above. But as the airstreams move away from their source regions, they are affected by conditions below them and by convergence, divergence, and mixing in the air ahead and above. In this way, their original characteristics are modified and eventually lost.

During winter, North America comes under the influence of three main air masses: polar continental, which originates in northern Canada; polar maritime from the northern Pacific; and tropical maritime from the Gulf of Mexico. *Polar-continental* air is bitterly cold but mainly dry, except in the immediate lee of the Great Lakes or where it is forced to rise, say, over the Appalachians. *Polar-maritime* air brings a great deal of rain to the western ranges of the Rocky Mountains as far south as the Sierra Nevada of California. But after crossing the mountains the air is much drier, and as it descends it is heated by adiabatic warming. Such dry, hot winds, common in spring, are known as the Chinook. *Tropical-maritime* air is warm and humid, and produces rain where it is forced to rise over mountains, or over a wedge of colder, denser air. Hot, humid tropical-maritime air gives very heavy showers and thunder-storms to the southern and central United States.

Britain is surrounded by air-mass source regions. In winter, polar-maritime air from northern North America and Greenland is warmed over the northern Atlantic and arrives in Britain as a cool, unstable but not particularly moist airstream that produces showers of rain and snow. Air drawn around the rear of a depression may reach England from the south-west after a long southerly journey over the Atlantic; such an airstream, known as *returning* polar maritime (rmP), has a low stable layer about 1500 feet deep, containing stratocumulus cloud. Higher up, the air is unstable with cumulus cloud forming over high ground. In summer, severe storms sometimes occur when the clouds and inversion of returning polar-maritime air are broken by rising thermals of warm air.

Colder, more unstable air from the arctic is called *arctic maritime*; it brings snow-showers and drifting snow to northern and north-eastern counties, particularly over high ground. Polar-continental air brings Britain its severest winter weather. When it comes in the form of easterly winds from an anticyclone over eastern Europe and northern Russia, temperatures hover around freezing point. They sink even further when south-easterly winds blow in from the frozen surfaces of central Europe. When these easterly airstreams travel across the North Sea they pick up moisture, so that snow-showers fall from stratocumulus clouds over the east coast, but these soon die out further inland. Polar-continental air in summer is warm and

dry, although it may produce fog and low stratus cloud over the east coast after crossing the cool North Sea.

Tropical-maritime air reaching Great Britain has its source in the semi-permanent anticyclone over the Azores. It arrives as a warm, humid south-westerly wind, with low cloud or fog trapped below an inversion layer. This inversion, produced by sinking and the movement of cold air near the ground, is generally at about 3000 feet; above it, the air is clear and dry. In winter, tropical-maritime air is associated with overcast, muggy days with drizzle and hill fog. Tropical-maritime air is warmer in summer than in winter, and daytime heating inland is then strong enough to break up the stratus cloud. The airstream is then unstable, resulting in heavy rain-showers, sometimes with hail and thunder, particularly in south-east England. *Tropical-continental* air from North Africa is a rare visitor to Britain. As a hot, dusty, unstable but dominantly dry airstream, it is responsible for some memorable, if usually short-lived, heat-waves.

When air masses of different temperature and humidity come together, they tend to remain separated by a boundary zone or *front*. Fronts (which were originally studied in detail by Norwegian scientist V. Bjerknes together with his son J. Bjerknes and other colleagues between 1910 and 1920) were first thought to be precipitous breaks in density. Today they are regarded rather as zones of very strong temperature gradients produced by the slow mixing of adjacent air masses; they are also much more complex in structure than was originally supposed.

When two liquids of different densities that do not mix are poured together into a container, the lighter tends to lie above the heavier, separated from one another by a horizontal discontinuity. In the atmosphere, warm, low-density air and cold, high-density air are also separated by a discontinuity, though this *frontal zone*, as it is called, is not horizontal but sloping, so the colder air lies like a thin wedge below the warmer. The slope of the frontal zone is very shallow, generally between 1 in 25 and 1 in 300. (An average slope in mid-latitudes would be

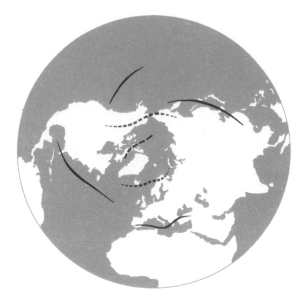

The principal frontal zones are the areas between the main air-mass sources. Map above shows the main frontal zones in winter in the Northern Hemisphere. Solid red line is the polar front, broken red the arctic front, and black the Mediterranean front.

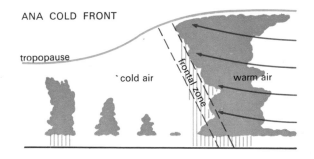

ANA COLD FRONT

Ana cold fronts (above) occur mostly during the winter in Britain, but are not common. The arrows indicate the movement of air in relation to the front, but this front is also moving itself—the cold air is advancing (from left to right).

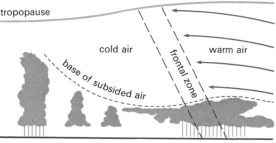

KATA COLD FRONT

In the kata cold front, the cold air pushes under the warmer air—as in the ana front—but the warm air tends to sink in relation to the front. A cold front is relatively steep, between 1 in 25 and 1 in 150, and the system may be only 100 to 200 miles wide.

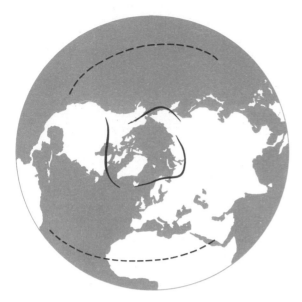

The map above shows the chief frontal zones in the Northern Hemisphere in the summer months. During summer, only arctic frontal zones are fairly constant. The broken line on the map represents the intertropical convergence.

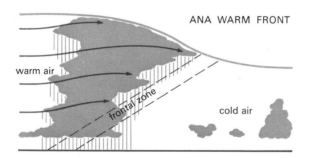

ANA WARM FRONT

warm air

frontal zone

cold air

Warm fronts (above and below) have a much gentler inclination than cold fronts, usually between 1 in 100 and 1 in 300. The weather resulting from such fronts thus affects a wide belt, possibly up to 600 miles across.

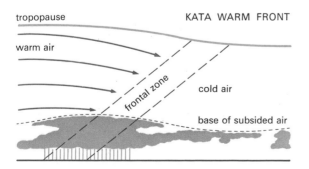

tropopause

KATA WARM FRONT

warm air

frontal zone

cold air

base of subsided air

As the warm air mass advances and climbs above the cold air, stratiform clouds are produced. The rain from a warm front is persistent, but not usually heavy; fog sometimes forms in the shallow layer of cold air.

about 1 in 100 so that at the top of the tropopause it would be 500 miles on the cold side of its position at the ground.) The breadth of the frontal zone increases with time, but it is usually about 1 mile deep, and its horizontal width varies from 25 to 300 miles. The fronts drawn on weather charts represent only the general position of these zones of mixing between air masses of different temperature and humidity. The frontal symbol is, in fact, generally placed on the warm-air side.

There are several recognized frontal zones in the patterns of world climate, although none is continuous around either hemisphere. They are found mainly over the oceans where there is less intense mixing of individually more uniform air masses. The *arctic front* lies at the junction of arctic and polar-maritime or polar-continental air. It occurs mainly in winter over the northern Atlantic and Pacific oceans. In summer, it is much weaker and exists only in very high latitudes. The *polar front* is found in lower latitudes at the junction between polar-maritime or polar-continental air and tropical-maritime air over the northern and southern Pacific and Atlantic and the intervening continents. It is most intense and lies nearest the equator in winter. The *Mediterranean front* forms over the Mediterranean Sea in winter where polar-continental air from Europe meets tropical-continental air from North Africa. The *intertropical front* or *convergence* (which was mentioned in Chapter 4) occurs at the meeting point of the trade winds of the two hemispheres and coincides with a belt of instability and heavy rains.

The movement of fronts and their related air masses is controlled by pressure and winds. At a *cold front,* warm air is displaced by colder air; at a *warm front,* cold air is replaced by warmer air. Usually winds change direction across a front. In the Northern Hemisphere, the wind veers—that is, its direction moves progressively clockwise in rough proportion to the intensity of the front. The considerable vertical air movement in the area of fronts is the controlling feature of frontal weather. If the warm air rises in relation to the frontal zone, the fronts are known as *ana-fronts*; if it sinks, they are

105

called *kata-fronts*. In practice, fronts may be both ana- and kata- at different levels. This complex structure when added to an equally complex history (which may include a reversal in direction) produces enormous variations in form. Even so, it is helpful to define these four standard types of fronts in temperate latitudes, while at the same time stressing that there is no such thing as typical weather at a front; every front is unique.

Like fronts, frontal depressions were first described in detail by J. Bjerknes and his collaborators in Norway soon after the end of World War I. And even after half a century, his original diagrams still appear in countless books on meteorology as striking evidence of his accuracy and insight. However, later findings—particularly observations of the upper stratosphere—have to some extent modified Bjerknes's original ideas.

The sequence of events leading to frontal depressions, originally described by Bjerknes and later modified by Tor Bergeron and others, centres on the growth and development of waves on the polar front. These waves corrugate the frontal zone like ripples on a water surface above a shelving coast. Water waves are, however, due to wind-shear at the discontinuity, while many more factors are involved in the case of fronts—factors such as the rotation of the earth and wind circulations set up by differences in density. The details of cyclone development remain something of a mystery; only recently has even a part of the puzzle been pieced together. But before we look at a few of the answers (as far as we know them), we need to describe the events that precede the growth and later decay of frontal depressions.

Let us take then the polar front, which usually runs from south-west to north-east across the Atlantic (and Pacific) and separates polar-continental (cP) and polar-maritime (mP) air to the north-west from tropical-maritime (mT) air to the south-east. Sometimes the cold and warm air masses are travelling in opposite directions, but generally both air masses are moving from the south-west. In the front, usually in its western part, a small local bulge or wave of warm air develops, and pushes into the colder air.

The wave then moves along the front at approximately the same speed as the warm airstream, while pressure falls a few millibars near the apex of the wave. If the wave remains shallow, it may travel as much as 600 miles along the front in a single day. This is known as a *stable* wave. *Unstable* waves, more than 300 miles long, develop in both size and wave-length as the associated low-pressure system deepens. The wave soon becomes wedge-shaped with warm and cold fronts marking the leading and rear edges of a warm-sector depression. Winds strengthen and back a little—that is, they blow in a progressively anti-clockwise direction: from south-west through south to south-east, for example. The depression itself increases in speed and follows a curving path, which also runs in an anti-clockwise direction in the Northern Hemisphere. The path of the depression follows very closely the direction of the isotherms (lines of equal mean temperature) between ground level and about 18,000 feet. These generally curve around from west through south-west/north-east to south/north as the air-mass pattern of the depression grows, matures, and decays.

Since the cold front in the rear of the depression moves faster than the leading warm front, the warm sector shrinks as its warm air is lifted over the cold air ahead of the warm front. The cold front first catches up with the warm near the tip of the warm sector to form a weaker front known as an *occlusion,* which separates polar-maritime (mP) air behind and returning polar-maritime (rmP) air ahead. If the rmP air is the warmer of the two, as is likely, the mP air cuts under it and forms a *cold occlusion*. Less often, except in winter when cold polar continental (cP) lies over Europe, the air behind the occlusion is warmer than the air ahead. It then rides over it to form a *warm occlusion*. Most of the depressions that travel across Britain are at least partly occluded; the southern half of the country is crossed by roughly an equal number of cold, warm, and occluded fronts. An occlusion is usually a rather weak front, often producing nothing more than an increase of cloud and perhaps a light shower; the depression itself, on the

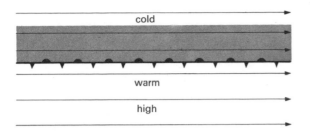

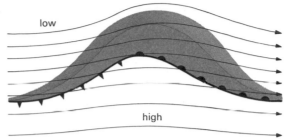

Above, a cyclone frequently forms when a warm air mass is adjacent to a cold one, with wind-shear at the discontinuity. Usually both air masses are moving in the same direction, but the effect is similar if they are travelling in opposite directions.

A small bulge of warm air develops, and begins to push into the colder air. An area of slightly lower pressure forms near the wave's apex; and once the bulge has begun to form, it frequently develops quite quickly into a sizeable one.

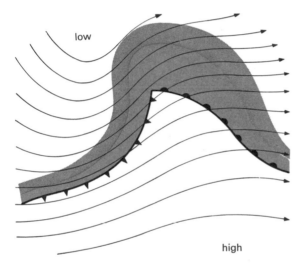

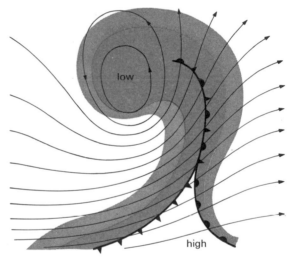

Above, the wave becomes wedge-shaped, and the rear cold front (left) gradually gains on the slower-moving warm front (right). The cold air pushes the warmer air upward, and clouds form, resulting in snow or rain.

Above, when the two fronts coincide, the warm air is displaced upward, and an occlusion is formed. Air rotates (anti-clockwise in the Northern Hemisphere) round the low-pressure area (centre top). A cyclone may have a diameter of 1000 miles.

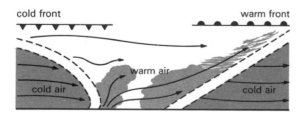

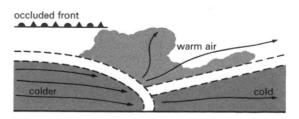

This drawing of a section through a depression (travelling from left to right) shows the cold air at the rear, and the warm air being lifted above the cold air ahead of the warm front.

Air following a depression is often much colder than the air in front of the warm front. This colder air tends to push in below the warm front, lifting the warm air up over the less cold air (right); this is called a cold occlusion.

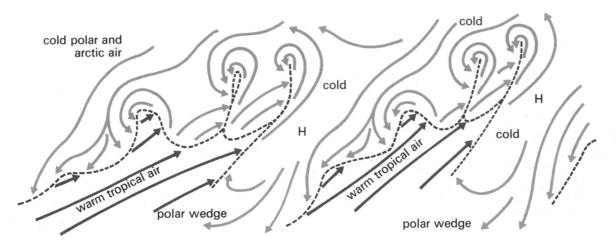

cold polar and arctic air

warm tropical air

polar wedge

cold

cold

H

warm tropical air

polar wedge

cold

cold

H

Often, from three to as many as six cyclones occur in a series, known as a cyclone family. Probably the leading cyclone will be occluded, and the second partly occluded; these will be followed by one or more waves about to develop into cyclones. As the leading cyclone dies out, new ones join on the end of the family. Such a cyclone family brings unsettled weather; the rain and strong winds typical of cyclones are interspersed with bright intervals accompanying the high-pressure ridges between the depressions.

other hand, now distinctly warm-centred, continues to deepen, with resulting strong winds. After a short time, however, the depression slows down, often partly because of a blocking anticyclone ahead (p. 114). Its speed and direction become erratic, and the winds move more and more obliquely across the isobars toward the centre of the depression, which finally fills and disappears.

We have already described the type of weather to be expected in the vicinity of fronts. Near the centre of a mature depression, the warm front is usually an ana-front with rising air; the cold front is normally a kata-front with (surprisingly to most people) sinking air. Farther away from the centre, these characteristics are frequently reversed. In this case, amounts of cloud and rain are greatest at the centre on the warm front, and on the more outlying parts of the cold front. An inactive warm front passing overhead is likely to be followed by an active cold front and vice versa.

In the warm sector of a depression, the slowly rising air leads to wide-spread though shallow stratus and stratocumulus cloud, which may produce light drizzle near the centre of the depression. Scattered cumulus clouds form in the polar air, and cumulo-nimbus in the rear of the depression. Depressions are generally most intense and travel farthest south in winter. Within 24 hours, winds may reach gale force, change direction by 180°, and die down to near calm.

Generally, the polar front is indented by a series or family of four, five, or more waves, all at different stages of development, though they usually increase in maturity from west to east. Such a series of depressions may be linked to one another by their fronts: the warm tropical air projects in a series of bulges and wedges into and above the cold polar air to the north. Often, especially in the west of the frontal system, the depressions are broken and destroyed by the polar air bursting through into the sub-tropical zone. The pattern may also be complicated by small, usually stable waves moving from west to east along a trail of rapidly forming and moving cold and warm fronts. Since these cold-front and warm-front waves, collectively known as *break-away* waves, are

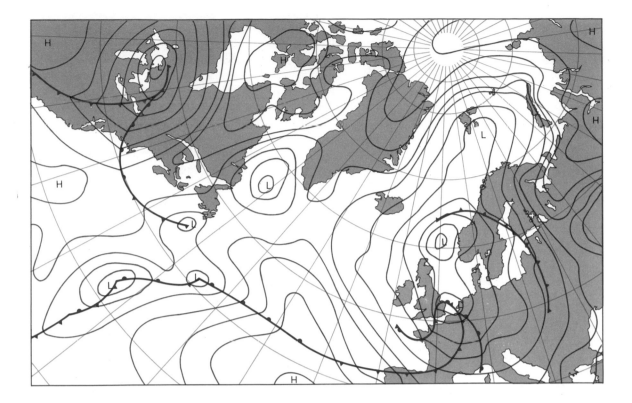

frequently associated with cloud and rain, they are of considerable interest to forecasters.

Most depressions in mid-latitudes form on the polar front, but there are some that are non-frontal. These are composed of a single air mass, and vary in size from the tiny, shallow depressions that occur over East Anglia, England, on a hot summer's day to the wide-spread summer lows that form over a whole continent. They are partly produced by variations in daytime or summer temperatures and pressures over land and sea, and are known as *thermal lows*. The largest—such as those that form off northern India in summer—are major features of the circulation; the smallest are weak and ephemeral, lasting less than 24 hours.

*Polar-air depressions* are also non-frontal. They form by local heating in an unstable polar airstream moving over warmer seas and sometimes over land. Because the heating is prolonged, they are deeper and longer-lasting than all but the largest thermal lows; they also occur in winter rather than in summer. Often they appear in the rear of mature or occluding polar-front depressions

Above, a typical synoptic chart, somewhat simplified, showing an actual cyclone family. Below, a chart showing no less than four cyclone families, at sea level. Such cylcone families are a frequent occurrence over the oceans of the world. They also form above land masses, though less frequently.

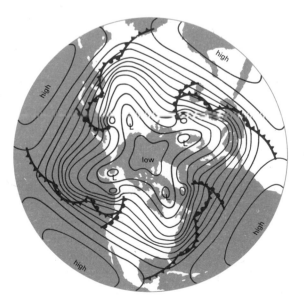

to form a *secondary* low that revolves with the air around the original depression, which is then known as the primary.

*Lee depressions,* sometimes known as *orographic lows,* are another type of non-frontal depression. They are found lying across the airstream in the lee of high mountain ranges such as the Rockies, the New Zealand Alps, and the French and Italian Alps. Frontal depressions may also form in these areas as fronts become steeper during their slow journey across a rough, serrated land surface, or because of air-mass contrasts on either side of the mountain range. They originate from the convergence of air in the lee of mountains, and can be compared to whirlpools downstream of a big stone in a river bed. For example, over the North Italian Plain, so-called *Genoa depressions* often form in the north-westerly airstream that crosses the Alps in winter. These depressions result in cool, showery weather in northern Italy and the more eastern areas to which they travel.

Surface lows are associated with (and are often caused by) the divergence of higher air, which often occurs ahead of the axis of a trough in the middle and upper troposphere. If the divergence is stronger than a compensating convergence near the surface, the depression deepens. Another important factor is the instability of the air. If the near-surface convergence is superimposed on a frontal zone, the front buckles and local wind systems develop (as described earlier) that tend to deepen the disturbance. Later, the depression may grow so high, perhaps even into the lower stratosphere, that it interferes with the divergence higher up, and so with the fall of pressure below. This is the mature stage of the depression. The final filling, occlusion, and decline of a depression is often caused by the advance of the upper-air trough so that convergence in its western limb becomes superimposed on the depression, which then disappears as a surface feature in three to six days.

The huge travelling depressions of mid-latitudes, an average of 1000 miles across, have no counterpart in the tropics. These areas, however, have their own kind of disturbance. One of these, the tropical revolving storm, produces some of the severest weather in the world. Tropical storms are known as *hurricanes* in the West Indies and southern United States, *cyclones* in the Indian Ocean, *typhoons* in the China Sea, and *Willy-Willies* off northern Australia. They appear on synoptic charts as very deep lows, 50 to 500 miles across, made up of almost perfectly circular isobars. Not surprisingly, before the use of weather radar little was known in detail about the structure of these frighteningly violent storms, which can toss ships and

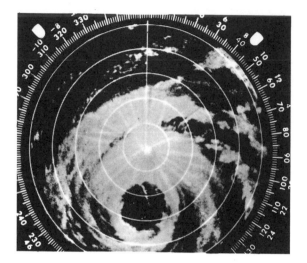

A radarscope picture of the hurricane Betsy, which swept across America in 1965. The picture clearly shows the shape of a typical hurricane, with a clear windless "eye" surrounded by high-speed winds and turbulent clouds.

Opposite above, the left-hand diagram shows the basic shape of a typical cyclone, with the characteristic area of low pressure at its centre. The right-hand diagram shows the movement of air (black arrows), flowing inward at the bottom, and out of the cyclone at the top. The low-pressure area pulls down warmer air from the stratosphere, thus forming the typical warm eye.

Right, wild seas sweep inland at Miami Beach, during a hurricane off the coast of Florida in September 1947. Such high seas, coupled with torrential rain and winds of 200 mph and above, have been the cause of much death and destruction.

aircraft around like corks in a nightmare of winds, rain, hail, and lightning. On land, tropical storms can devastate crops and buildings with shrieking winds, flood river valleys with deluges of rain, and swamp coastal areas with huge sea waves driven before the storm.

Tropical storms reach from top to bottom of the troposphere. The rapid convergence of the air spiralling inward below about 10,000 feet would soon fill the low-pressure centre in the absence of equally intense ascent and divergence (in an anticyclonic system) between 10,000 and 30,000 feet. At these heights, pressure gradients are weaker than at the ground, but the strong centrifugal force leads to a quick spiralling divergence of the rising air. This divergence is balanced by the subsidence of high tropospheric and lower stratospheric air in the centre, or "eye," of the storm. Near the ground, the eye is a warm, dry, cloudless region with light winds or an eerie calm, and temperatures several degrees warmer than those of the surrounding storm.

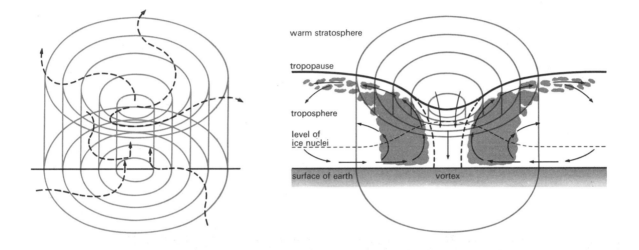

This central calm, no more than 10 to 30 miles across, is surrounded by an amphitheatre of cumulonimbus clouds from which falls a deluge of rain accompanied by winds of up to 200 miles an hour and occasionally more. Wind and rain are particularly strong within 100 miles of the centre. Here, 5-inch falls of rain in 24 hours are common, and 10 inches far from unusual. Near the top of the storm, a ring of crystalline cumulonimbus anvils marks the edge of the storm.

Tropical storms are now tracked quite easily by radar and weather satellites; the task of a warning system is also made simpler by the fairly regular movement of storms around the western ends of sub-tropical anticyclones in the seas of both hemispheres. Only the southern Atlantic is free of such storms—probably for reasons connected with the relative coolness of its sea surface.

Most storms lose their ferocity on leaving the tropics, but occasionally they penetrate deeply and destructively into extra-tropical areas. In the autumn of 1954, for example, hurricanes "Carol," "Edna," and "Hazel" devastated large areas of the eastern United States and Canada. Some of the less violent extra-tropical invaders are given a new lease of life by high-level divergence, or even air-mass injections in temperate latitudes, though they always betray their place of origin by the circular form of their isobars.

The origin of tropical storms is still something of a mystery. Most recent theories discount the old explanation based on waves on the intertropical front. The most widely accepted explanation is that of the American meteorologist Joanne Malkus, who suggested that hurricanes develop in a region of convergence and deep instability behind shallow

Below left, looking across the warm eye of a hurricane (broken cloud) to the more solid ring or wall of rising air and cloud surrounding the eye. Below right, the eye of a hurricane photographed from a Mercury weather satellite at about 100 miles above the earth's surface.

Right, a computer's interpretation of a storm. A weather satellite supplies an ordinary photograph of a storm; this photo is then scanned electronically and its various shades of grey converted into letters and symbols as shown here. A computer can make use of these letters and symbols later.

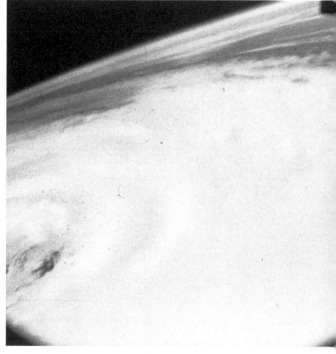

wave disturbances that move slowly westward through the trades. Occasionally, the wave breaks the trade-wind inversion that normally limits the clouds of the trade-wind belt to heights of between 6000 and 10,000 feet. The easterly wave may then deepen and form a closed, low-pressure vortex, especially when the wave penetrates a high-level anticyclonic circulation with rapid divergence so that the cyclonic circulation is intensified near the ground. Such disturbances frequently result in gales and heavy rain-showers, but only a few deepen enough to produce true hurricanes. These probably occur when a warm "eye" is formed by descending air in the centre of the storm, but so far nobody really knows why or how this happens.

From depressions, we turn now to anticyclones—areas of high pressure that is highest in the centre and around which winds in the Northern Hemisphere circulate in a clockwise direction (anti-clockwise in the Southern). Because of the balance of forces that affect the winds, anticyclones are never as small or as intense as most depressions. Generally, they are several thousand miles wide, with light (and near the centre, variable) winds. Again, unlike depressions, anticyclones usually originate as extensions of existing highs rather than as independent systems. Anticyclones are also normally accompanied at low level by cooling, whereas depressions are accompanied by warming.

In anticyclones, the air sinking slowly down to the turbulent lower troposphere is compressed, and so becomes warmer and drier. Even so, the weather produced by anticyclones and smaller ridges of high pressure is also affected by other factors such as the prevailing air mass, the season, and the time of day. In summer, they are responsible for spells of fine, warm, or even hot weather, and showers, some of which are heavy, with hail and thunder. In winter, they may give bright, cold, crisp weather or cold, overcast weather with heavy snowfalls or persistent radiation fog. The association of high pressure with fine weather assumed by the traditional weather glass is often an oversimplification, since anticyclones give persistent, but not necessarily fine, weather.

Anticyclones must be distinguished into cold- and warm-centred systems. In cold anticyclones, pressure falls off rapidly in the dense air, so they seldom rise above 10,000 feet. In north-polar anticyclones during summer, the high is small, shallow, and short-lived, with advection sea fog or low stratus cloud below the inversion layer. In winter, this high-pressure area stretches further south with separate centres over North America, Greenland, and Russia. These give generally clear skies and very low temperatures—often as low as 50°C of frost. In the mid-latitude, westerly belt, depressions are normally separated by ridges of high pressure rather than by anticyclones. These frequently result in a one- to two-day interval of bright weather between the periods of disturbed weather associated with the depression. By day, skies are broken with low, scattered cumulus clouds trapped beneath the inversion; by night, clear skies and light winds cause temperatures to fall with the possibility of fog, dew, and ground and air frost.

Warm anticyclones are much deeper than cold ones because of their low air density; they often reach well into the stratosphere, usually as a ridge rather than a circular high-pressure system. The anticyclones that circle each hemisphere in sub-tropical latitudes are typical warm highs. They expand and contract rather than move, and result in wide-spread light winds and scattered cumulus showers. Warm anticyclones play an important part in the weather of enormous areas of the world's surface; their size and depth also make them important factors in the pattern of the general circulation.

When sub-tropical highs and the mid-latitude depressions on their poleward sides have abnormally long west-east axes, the circulation also moves mainly in a west-east direction, and is known as *zonal*; when, on the other hand, their alignment is clearly north-south, the circulation is said to be *meridional*. This situation occurs when several deep, warm anticyclones are centred in mid-latitudes, say about 55°N, blocking the west-east currents and steering depressions much further north or south than usual. *Blocking highs,* as these anticyclones are called, often form at low level below a detached extremity of an over-developed ridge of warm air in the middle troposphere. The zonal (west-east) airstream then re-forms in lower latitudes to isolate a warm blocking high astride the chain of moving depressions in mid-latitudes. At the same time, a pool of cold air may break away from the tip of a neighbouring, attenuated cool trough. This appears on the surface chart as a *cut-off depression*. In either case, the detached anticyclone or depression often slows down to a standstill or may even drift slowly to the west. Blocking highs frequently form over Europe, where they often play a vital part in controlling the weather. Even a small shift in their position may substitute a dominantly anticyclonic type of weather for the quite different conditions brought by a bank of depressions following one another slowly around the margins of the high. They may even tip the scale between a notably dry and warm summer and one particularly remembered for its overcast skies and rain. Even in one year, south-east England can enjoy the first type of summer and north-west Scotland suffer the second.

Anticyclones form in a way that is almost the reverse of the process leading to depressions—that is, by convergence at high level followed by subsidence that is not quite levelled out by divergence below. In cold highs, the convergence and sinking result from contraction and loss of density in tropospheric air. The high-level convergence in warm anticyclones is more difficult to explain. It probably occurs on the leading edge of a slow-moving upper-air ridge because of changes from anticyclonic to cyclonic curvature. These in turn alter the direction of the centrifugal term in the balance of forces on air moving through the wave.

With this examination of depressions and anticyclones, we have completed our survey of the various air movements that determine the weather over areas extending for hundreds or even thousands of miles. Such distances provide the scale for the 24-hour weather forecasts and the synoptic charts on which they are based. In the next chapter, the scale will be reduced as we examine some of the factors that, by modifying regional weather conditions, are responsible for creating local climates.

In winter, an anticyclone may give rise to persistent dense fog, such as the London "pea-souper" above. Such weather may cause many deaths, through road accidents and illnesses.

Fair-weather cumulus cloud whose vertical development is limited by the subsidence of air in an anticyclone. Contrast this with the rain-bearing cumulus clouds on page 103.

# 10 Local Climates

The term *local climate* is generally applied to changes in weather over short distances—the variations, for example, between the bottoms of valleys and the tops of hills, or between town and country. On a smaller scale still, climates can differ from one side to the other of a hedge or garden wall, or from the base of a bush to its leafy crown. Such conditions are usually described as *microclimates*. In this chapter, we shall be looking at some of the local factors that make up the study of climate on this scale—factors that generally assert themselves most strongly during calm, clear, anticyclonic weather.

The shape of the land has an important bearing on local climate for several reasons. To begin with, mountains thrust their heads above the low-lying blanket of cloud, dust, and smoke into the cleaner, more rarefied air above. On mountains, then, there is more short-wave ultra-violet radiation and less scattering of sunlight into shaded areas. One therefore acquires a suntan very easily; there is also a sharp drop in surface temperature when one moves out of the sunshine into the shade. Because of the horizontal movement of cool air from high above the surrounding lowlands, the daytime air is usually cool even though soils and rocks that are heated by the sun may be very warm. In fact, air temperatures decrease with height at a very variable rate. As the British climatologist Gordon Manley has shown, in the highlands of western Britain, the fall is very rapid, averaging about 1°C per 500 feet, so that the highest point at which trees can grow—the *climatic tree-line*—is reached at about 2200 feet. In practice, however, trees are unable to grow well below this line because of strong winds and saturated blanket bog that hamper their growth on high ground. Near the coast, sea spray creates further difficulties, so that trees do not flourish above 1000 feet, though a few isolated trees may be found up to 2000 feet in sheltered pockets. Above the tree-line in Britain, there is a tundra-type vegetation with mosses and bogs at altitudes that could be cultivated in central Europe.

The temperature in upland areas is not controlled by altitude only. Often there is a sharp difference in temperature between the windward and leeward or between the sunny or shaded sides of hills, and always between day and night. At night, cold, heavy air drains off the slopes and accumulates in the valleys or behind some obstacle such as a hedge or embankment. Such cold-air drainage occurs in every valley, but sometimes it is so pronounced that the valley acquires a reputation as a *frost hollow*. In England, the best-known example is the small valley of the River Chess, which cuts into the back slope of the Chiltern Hills north-west of London. The dimensions of the valley are far from unusual; it is only a few hundred feet deep, and its counterpart can be found in almost any stretch of rolling country, while much more exaggerated examples occur in high mountain areas. Yet this modest valley has one of the severest winter climates ever recorded in Britain. No month of the year is free from the risk of air frost, which occurs, on average, on two nights out of five; ground frost is even more frequent, occurring on three nights

A deep valley, such as this one near Engelberg in
the Swiss Alps, has a pronounced effect on local
climate and, consequently, land use. The floor of
the valley is fertile farmland; above this is an
area of forest; above the tree-line, grass pasture
(snow-covered in this photograph); and still higher,
an inhospitable region of rock and snow. A
valley's direction is also important. Many crops
needing a fairly warm climate, such as grapes,
can do well on the sunny side of a sheltered
valley; on the colder, shaded side of the valley,
however, such crops cannot be grown.

out of five. The cold air accumulates from the surrounding slopes until it fills the bottom of the valley to a depth of 30 to 40 feet. In such valleys, night temperatures are often 3-4 and sometimes 6-7°C lower than those on adjacent slopes and hill-tops. In some deep, snow-covered mountain valleys, differences three times greater have been recorded.

Obviously valley bottoms should be given a wide berth by the farmer who wants to avoid crop-damaging frosts, and by the householder who is anxious to keep down his heating costs. The best sites are those half-way up a slope—above the clear night frosts but below the generally cool and windy hill-top.

Anyone choosing a site for a house or farm should also consider the aspect. Clearly, slopes that face the sun, especially the afternoon sun, receive more radiation than those that do not—particularly northern slopes, which may be in perpetual shadow. In the deep valleys of high mountains like the Alps and Rocky Mountains, very little light is scattered on to slopes that face away from the sun. They are therefore much cooler than those facing south. In many mountain valleys, the villages and cultivated fields are sited only on the sunny side.

The distribution of humidity in hilly areas is very complicated; it depends on such factors as altitude, wind direction, and the prevailing airstream. In general, relative humidities are highest in the frequently fog-shrouded valleys in winter and on the often cloud-topped hill-crests in summer. But the most persistent contrast between hill and vale is the difference in wind strength between breezy hill-tops and sheltered valleys. In addition, winds in the lowlands are strongest by day while those in uplands are strongest by night. This is because thermal turbulence by day transfers kinetic energy from the high, fast-moving air to the air at lower levels, while at night there is little exchange of energy between different heights, and air near the ground is slowed down by frictional drag with the earth. Obstacles in the path of the wind of course provide considerable protection. Trees and plants that would normally never survive in upland areas may grow in the shelter of a walled garden. Such shelter not only reduces physical damage but increases temperatures, partly by cutting down the mixing of warm air near the ground with higher, cooler air, and partly by the reflection and radiation of sunlight from the walls. In this way, peaches and grapes can grow on south-facing walls surprisingly well in northern areas.

As a general rule, wind speeds in upland areas increase with height, just as they do in the free atmosphere. But as air approaches a mountain, it is forced to concentrate its flow and so increases its speed near the crest in the same way as a river rushes more rapidly over a weir. But in the lee of hills, the wind drops in speed and eddies are formed as air moving toward and up the lee slopes meets the regional wind travelling in the opposite direction at a higher level. So someone standing on the top of a steep hill and looking across a valley can often see smoke climbing up the lee slope, even though the wind above the hill is blowing in the opposite direction. In the lee of hills that slope more gently down to a wide plain or valley (particularly when the airstream contains a high-level inversion), winds travel down (not up) the hill. Often an eddy forms some way down-wind of the hill-top, which lies below a stationary or lee wave with a cloud forming in the crest of the wave and at the top of the eddy.

Winds, again like water, also increase in speed as they are channelled through gaps and along valleys. Mountain-gap winds are not only strong and gusty but they are often unusually warm or cold. The cold winds that sweep through the Cajon Pass, which connects the Mojave Desert and the San Bernardino Valley in southern California, have created a break in the fruit orchards opposite the pass's entrance by bringing freezing weather in winter and beating the trees about in summer. The *Mistral* is another torrent of bitterly cold, dry air that flows through the narrow passage of the Rhône valley as northerly winds blow over southeast France from high pressure over northern Europe to low over the Ligurian Sea.

One type of local wind takes its name from the *Foehn* winds of the northern foot-hills of the European Alps, particularly in

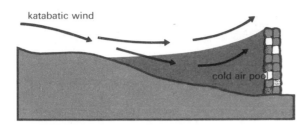

katabatic wind

cold air pool

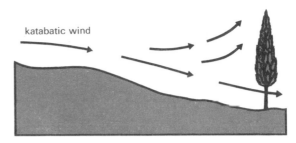

katabatic wind

Above, cold air tends to drain downhill to low-lying land. If a solid wall or fence is built across a slope, it dams back a pocket of cold air. To prevent this, it is advisable to use an open form of boundary. A row of trees, for example, will allow the cold air to drain between the tree trunks.

On the other hand, a walled garden—if correctly sited so that it does not trap cold air, and preferably facing south—can provide a sheltered environment. In such a situation, the peach trees (below) at East Malling Research Station, Kent, England, are flourishing. Vines can also be grown outdoors in southern England under such conditions.

Austria and Switzerland. The same kind of hot, dry wind is known as the *Chinook* on the eastern slopes of the Rocky Mountains and as the *Zonda* in the Andes foot-hills. Similar winds are also found in the lee of the New Zealand Alps and less well-developed examples even occur in the lee of such modest uplands as the high areas of Scotland and North Wales. Foehn winds are often heralded by a sharply defined cloud over the mountain tops, which is known as the *Foehn Wall* in Europe, and sometimes by one or more roll clouds in the higher, cooler crests of standing waves in the airstream to the lee of the mountains. A rise in temperature of as much as 27°C in two minutes has been recorded in the foot-hills of the Rockies, and increases of 33°C in 12 hours are quite common. These high temperatures are often associated with humidities as low as 10 or 20 per cent. Foehn winds are most frequent in late winter and spring, but may occur whenever air is drawn across the mountains. Often they start a rapid thaw of winter snow and increase the danger of forest fires in summer. They also seem to have a psychological effect, and have even been linked to sharp increases in local suicide rates. The hotness and dryness of the winds were originally thought to result from heat liberated during condensation, and moisture lost during precipitation on the windward slopes. The air therefore rises and cools at the saturated adiabatic lapse rate, but descends the lee slopes and warms at the higher dry adiabatic lapse rate. This explanation now seems inadequate, and it is thought that the winds' warmth and aridity is also due to the subsidence of air (sometimes from an inversion layer) that was originally at a higher level on the windward side as it moved above air banked up against the mountain barrier.

Hilly areas also generate local winds during calm weather. By day, hill slopes are heated by the sun and the air above them becomes warmer than the air at the same high level above the adjacent valley or plain. Because of this, pressure surfaces are distorted, and a circulation is set up with an *anabatic* or *valley wind* blowing gradually uphill complemented by a sinking in the free

A cold dry wind often blows off from the coast of Greenland. During the cold winter months, cold air gathers over the interior ice-cap plateau. The approach of a low-pressure system may draw a strong current of cold air out toward the sea, whipping up snow from the ground and thus causing blizzards. This type of wind is usually accompanied by a wall of cloud, which covers the land but stops abruptly at the coast-line. The example above was photographed with the royal yacht *Britannia* anchored off shore.

air above the plain. After sunset, the air over the hillside cools, and the wind circulation is reversed. Cool *katabatic* or *mountain winds* then flow downhill. The flow in both cases is far from simple, with branching and twisting air currents in the ascending daytime wind, and tributary airstreams moving down the valley sides to join the main flow along the valley by night. Even gentle slopes of only two degrees can start air moving on clear, calm nights, but the strongest flow occurs across steeper slopes draining a large catchment of air. Such night winds, which are normally shallow, can be as deep as 300 feet in the deep valleys of mountainous regions, but they rarely move faster than 10 mph unless they are concentrated through gaps or are supplemented by regional winds. Daytime valley or anabatic winds are deeper, reaching 800 feet in some areas, though their flow is more gentle and diffuse.

In hilly areas where moisture-laden winds blow mainly from one direction, the windward slopes are usually wetter than the leeward, with most rain falling near the crests of the hills. Increased rainfall over high ground is due to several factors. These include the compulsory ascent of air over the mountain; the increased instability within the airstreams, particularly in those where there is a rapid fall in humidity with height; local instability above warm surfaces; and the steepening of fronts. But while this pattern of more rain on windward slopes and less on their lee, or "rain-shadow," side is true of most highlands, the reverse situation is often found on low or isolated hills. Here there is often more rainfall on leeward slopes: first, because the large droplets that are held suspended by the strong updraughts to windward are released during the descent of the leeward slopes; and second, because of the leeward convergence and ascent of air moving around and above the hills. Another factor may be the time-lag between cloud formation and the growth of raindrops.

Having considered the shape of the land, we must now take a look at the composition of its surface, which has an important effect on climate. Different soils respond in different ways to heat and moisture. At one end

During the day, air above high slopes in the sun warms more than air at the same altitude above the valley. The result is an anabatic wind (below left), drawn up the hillside from the valley. At night, air above the high land cools more quickly than the air in the valley, and so a katabatic wind (below right) flows down the hillside into the valley; it is usually stronger than an anabatic wind, though rarely over 10 mph.

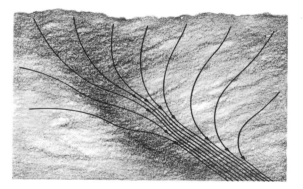

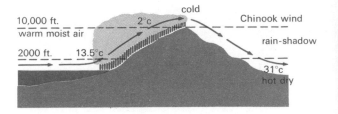

The hot dry Chinook wind, which blows from the Rocky Mountains, is still somewhat of a mystery. Warm, moist air from the Pacific Ocean, forced to rise when it meets the Rockies, cools and drops some of its moisture as rain on the western slopes. Then it descends on the east, warming about twice as quickly as it cooled on rising, but having lost much of its moisture.

of the scale, there are the bare rock surfaces that warm up quickly on a clear summer's day, especially if they are dark colours and reflect little radiation, like basalt or asphalt. Dry sandy soils have a higher albedo than dark clays, but because they contain so much air between the grains they have smaller thermal capacities and conductivities. Sandy soils therefore warm up quickly by day and during spring. For this reason, they are called "warm" soils by farmers, though they also cool rapidly on calm, clear nights and encourage ground frosts. Clay soils, which are generally more compact than sandy soils, retain more water, so that temperatures are less variable at the surface and heat penetrates more deeply. The conduction of heat to and from lower levels also helps to stabilize surface temperatures in clays and loams. Not surprisingly, well-dug clay behaves in much the same way as sandy soils. And wet soils tend to be cooler by day and warmer by night than dry ones, because of daytime evaporation and the ability of water to conduct and to hold heat. On clear nights, differences of as much as 8.4°C have been recorded (at four feet) above adjacent sandy and clay soils. Snow is the coldest surface of all, since it has an extremely high albedo and low ability to conduct heat. On the other hand, it insulates and protects the surface on which it lies and any crops that it covers.

The energy and moisture exchanges of surfaces are further complicated by vegetation. And since the vegetation rather than the soil is the surface that is exposed to the atmosphere, its form is very important. Grass and grass-like vegetation that trap a great deal of air behave very much like sandy soils —that is, there is a wide range of temperature within and immediately above them, although daytime temperatures are to some extent reduced by evaporation.

Vegetation with large crowns, such as trees, behaves rather differently. To begin with, it certainly reduces surface run-off by suspending rainfall in its crown so that it evaporates before reaching the ground. Forests also have low albedos with values as little as five per cent for pinewood and equatorial forests. Most of the sun's radiation is

Even a slight frost can make the difference between a good fruit crop and a failure, so a fruit farmer tries to combat frost. In this experimental orchard at East Malling, Kent, the ground under the fruit trees has been ploughed to keep it clear of vegetation. Such treatment prevents the temperature from falling quite as low as it would otherwise do, and the small difference may save the fruit crop.

absorbed by the upper leaves and branches of a thick forest, though the exact amount depends upon the type and closeness of the trees. A dense forest is refreshingly cool on a hot summer's day, and at night it stays comparatively warm; most heat is lost from the tops of trees, though cold air may spill through to the ground whenever this crown layer is broken. Wind speeds are of course reduced inside a forest; the form of the circulation is also changed. It has been estimated that in the case of a forest in full foliage, winds at a level of half the height of the trees drop to less than 10 per cent of their speeds in the open. Lower down, speeds are even more severely restricted, and near the ground, the air hardly moves at all, so that it is soon saturated by evaporation. Humidities are also high in the air between the leaves. Spongy, porous forest soils therefore tend to remain damp in spite of reduced rainfall.

Rainfall is not always reduced inside forests. Where woods are shrouded in frequent fogs or low clouds, rain gauges placed under trees catch more water than they do in the open. This is because the fog or cloud droplets are driven by moderate or fresh breezes against the trees where they collect and drip to the ground. This phenomenon, known as *fog drip*, often provides the main supply of water to the trees. It occurs, for example, in the forests on the Berkeley Hills overlooking San Francisco Bay. Here, pine and eucalyptus trees at about 1000 feet above sea level precipitate as much as 10 inches of fog drip during the dry but foggy summer months—an amount equivalent to nearly half the area's annual rainfall.

The effect of woodland on the actual amount of precipitation falling on to and around them is less certain. Forests add water vapour to the air, but, as we saw earlier, the main cause of precipitation is not the amount of water vapour in the air, but the existence of conditions that favour its release; of these, vertical motion and cooling are the most important. There is little doubt that air motion is affected by forests, but whether their influence is great enough to increase local precipitation is doubtful. The many investigations into the question

have failed to produce any definite proof either one way or the other, since almost all the experiments contain a possible flaw in measurement or analysis. This is not surprising, since it is extremely difficult to measure rainfall within and above forests, and equally hard to estimate or eliminate the effect of other factors such as local topography. Another problem is to decide whether such differences are not simply the result of changes in regional climate. There have of course been references throughout history to droughts following on forest clearance. But many of these probably resulted from increased run-off and soil erosion brought about by the removal of the protecting leaf canopy and the binding tree roots from the soils. In other words, such droughts were probably due more to the state of the land than to the weather.

From land surfaces, we move now to water surfaces, which affect climate in four ways: by acting as heat banks, since they are slow to warm and slow to cool; by transporting heat; by presenting a comparatively smooth surface; and by supplying most of the atmosphere's water vapour.

The heat capacity of water is between twice and three times that of most land surfaces. In other words, two to three times more heat is required to raise the temperature of a unit volume of water than is necessary to heat the same volume of solid earth by the same amount. More important still, only a thin layer of the earth's crust is affected by exchanges of heat: very often it is difficult to detect any daily variation in soil temperature below four feet, although the effect of the annual variation may reach a few feet deeper. In water, on the other hand, solar radiation penetrates relatively deeply and the stirring motions cause a fairly deep layer to be heated. For these reasons, land has a quicker and greater surface response to radiation than water. So areas inland have a wider daily and annual range of temperature than areas affected by winds from seas and large lakes. The seas around Britain are warmest in August or early September, and coolest in late February or March—in both cases a month or two later than the land. But the effect of seas and lakes depends not only on their response to heat but also on the way in which they circulate. Like air masses, moving warm and cold water currents affect the temperature and amounts of cloud, fog, and precipitation in neighbouring regions. For example, when the prevailing wind blows off the sea, the cool coastal waters of central California are mainly responsible for its cool, foggy summer weather.

How far this maritime influence penetrates inland depends upon the prevailing winds. In North America, where the prevailing wind is westerly, continental conditions creep very near to the eastern coastlands in winter, while a similar climate only begins far inland in the corresponding latitudes in Europe. Here topography is important, for the Rocky Mountains of North America prevent the sea from affecting more than a narrow coastal strip, while in Europe, there is little (except in Scandinavia) to block the path of the prevailing westerly winds.

Coastal areas are also affected by their position on the boundary between land and sea. High cliffs often trigger off instability in winds blowing in from the sea to form cumulus clouds, which then drift inland during unstable weather. In summer, skies are generally clearer over the sea and coast because of daytime subsidence above the relatively cool sea; as a result, there is more sunshine and less rain than inland. By night, storm-clouds may build up over the relatively warm sea, while the skies inland become clear. These changes are associated with land and sea breezes, the familiar daily reversal of local winds across coasts during otherwise calm weather. These breezes are complex in form, but all are the result of the unequal warming of adjacent land and sea. By day, air rises over the warm land and moves out to sea at a height of several hundred feet before sinking. This upper movement is complemented near the surface by a cool *sea breeze*. The circulation reverses soon after sunset, when the land cools faster than the sea. Air now sinks over the land and rises over the sea with a low-level linking flow of air from land to sea, called a *land breeze*. Because temperature differences are usually greater by day

than by night, the sea breeze is normally much stronger than the land breeze. And since both are too localized to be affected by the Coriolis force, they first move almost directly from high to low pressure at right angles to the coast. In mid-latitudes, sea breezes seldom exceed 15 mph and are usually less than 500 feet deep; in the tropics, they are stronger and deeper. Their penetration inland depends on the direction of the regional wind. If it is blowing from land to sea, the colder sea breeze undercuts the warmer land air and makes the beaches and coasts rather cool. But only a few miles inland, the breeze becomes slower and warmer, so that only a narrow coastal strip is affected. When the regional or gradient wind blows in the same direction and so reinforces the sea breeze, the combined effect may be felt much farther inland—possibly for 50 miles in England or even more along large river estuaries such as the Thames and Severn.

Inland waters also affect local climates, though their effect is weaker than that of the sea and often masked by regional weather changes. Even so, certain large inland lakes, such as the Great Lakes in North America, have a considerable influence. The American climatologist H. E. Landsberg has studied the climates to windward and leeward of the Great Lakes and has shown that the leeward sides to the north and east of the Lakes have a generally warmer, wetter climate than those to windward. It has been estimated that the average minimum temperatures at Milwaukee and Grand Haven, which lie 80 miles apart on the opposite sides of Lake Michigan, differ by 5.6°c. Grand Haven, on the eastern shores of the lake, is not only warmer but has a longer frost-free season and slightly higher humidities.

These then are some of the factors that, by modifying regional conditions, produce local climates. Most of them are caused by natural phenomena such as hills, woods, and lakes, though forest clearance and drainage may have to some extent modified their original form. In the next chapter, we go on to look at a type of climate that is largely man-made—the changes and innovations created by the building of cities.

Seen from an aircraft, these cumulus clouds closely follow the outline of the coast below them. As cool, damp air passes over the relatively warm land, thermal uplift leads to the formation of cloud.

Day and night changes of temperature give rise to air movements across coasts because land warms and cools more rapidly than water. During the day (below left), land heats more rapidly than sea, and a pressure gradient is formed near the earth from land to sea that results in a cool sea breeze. At night, land cools more rapidly than sea, and a flow of air from land to sea—a land breeze—develops.

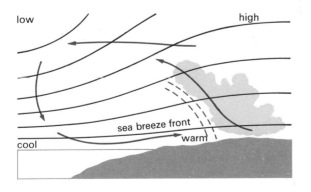

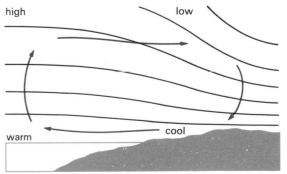

# 11  Town Climates

Just as cities represent man's most drastic re-shaping of the natural form of the land so urban weather provides some of the most dramatic examples of man-made changes of climate. The widespread replacement of fields and woods with houses, factories, and power-stations, and of country lanes with surfaced roads, has created a new and distinctive type of local climate—a *town climate*. In towns, both the chemical and physical make-up of the atmosphere is changed in a way that must strike even the most casual observer. Urban buildings are dirtier and their windows need to be cleaned more often than those of country houses; flowers bloom earlier and snow melts more quickly in city parks than in rural gardens. On autumn evenings, many surburban ring-roads are covered with mist, and in many central areas the early morning sun is masked by fog. On a still smaller scale, a whole range of micro-climates can be found within the same street.

In most large cities in temperate latitudes, the most striking departure from natural conditions is a shroud-like presence of solid, liquid, and gaseous pollutants. The term pollution of course covers mineral dusts from soils, quarries, and roads, and pollen from plants as well as the products of combustion.

It therefore occurs not only in cities. One of the most virulent pollutants, which causes allergies and hay-fever in the American Mid-west, is the pollen of ragweed. But the most damaging pollutants are associated with industrial and urban areas, where they are poured out in a bewildering variety from houses, factories, steel mills, incinerators, quarries, motor vehicles, railway engines, and ships. Sometimes pollutants are prevented from escaping by a warm inversion layer that, lying a few hundred feet above the chimneys and exhausts, traps and flattens the smoke plumes rising beneath it. Concentrations may become so heavy that lights are needed at midday. The black shroud eventually coats everything and everybody in a shower of smuts, which builds up to a thick black crust or fur. When combined with oxygen and water, pollutants can crumble masonry, corrode metal, dissolve nylon, make rubber and leather hard and brittle, and kill plants and animals, including man. The cost is colossal: a conservative estimate for Britain is £250 million a year, which amounts to £10 per head in the worst areas and £5 per head else-where—a figure that of course excludes the cost of the fuel.

A helpful way of looking at pollutants is to divide them into two classes: first, the smoke and sulphur dioxide released by burning bituminous coal and heavy fuel oil. The London and Pittsburgh atmospheres were the classic cases of this lethal carbon-sulphur mixture, especially when combined with fog droplets to produce smog. The second group consists of gaseous hydrocarbons, mainly from motor vehicle exhausts. Los Angeles provides the outstanding example of this type of pollution. These three cities are built in low-lying areas that are particularly prone to inversions. In rural areas of Britain, coal and oil are responsible for less than half the pollution, but in most towns the majority is produced by these two fuels. During any year in Britain, about one and a half million

Most large cities in temperate areas (such as New York, above) experience periods of smog. A warm inversion layer just above the buildings of the city traps smoke and fumes, and the pollution gradually builds up and mixes with a dense mass of fog that can bring the city virtually to a standstill.

The thick crust of dirt and soot that covers so many city buildings can destroy masonry and spoil fine carvings. Happily this panel on the outside of St. Paul's Cathedral (right) has not been so severely attacked. Possibly because its layer of grime (top picture) first began to collect in the days before harmful waste poisoned the air, the panel, believed to be the work of Grinling Gibbons, was revealed unharmed and in its full beauty (bottom picture) during the recent cleaning of the cathedral.

tons of smoke are emitted, of which four fifths come from domestic fires (mainly between September and April), though in London the figure is 90 per cent. Today however the output of smoke is decreasing as a result of statutory control and the gradual replacement of coal with coke, oil, and gas.

Among the gases produced by combustion, those containing sulphur are the most harmful. Sulphur is found in coal and in various fuel oils, and, although it can be removed from flue gases, the present cost is uneconomic. Since the changeover from coal to oil in Britain following the Clean Air Act of 1965, emissions of sulphur dioxide into the air above Britain have increased as the output of visible smoke has decreased and the total is now estimated at over six million tons per annum. New York alone produces about one and a quarter million tons annually plus two million tons of carbon monoxide from petrol exhausts. But the city is reasonably well ventilated, so the effect is less severe. In strong sunlight, sulphur dioxide combines with water to produce sulphuric acid, which has obvious ill effects on both men and

materials. Nitrogen oxides also occur in diesel fumes, and some of the more noxious secondary products are ozone and the class of organic compounds known as aldehydes— toxic gases that are held responsible for the eye irritation in such places as Los Angeles.

Urban air contains many times more suspended particles or aerosols than elsewhere. This suspended material is of course eventually deposited, but where and when depends upon the mass of the individual particles, the height, speed, and temperature at which they are exhausted, and the vertical distribution of temperature and wind. The largest dust and ash particles fall to the ground near their source, but the lighter smoke particles may be carried more than 100 miles in certain conditions. Rates of deposit are sometimes staggeringly high: during the 1950s some densely settled, low-lying districts of London were coated in more than 400 tons per square mile per year, and most suburbs in more than 200 tons. The whole of London collected more than 100,000 tons annually. In some areas of heavy industry, a deposit of 1000 tons a square mile every year is not uncommon,

A warm inversion layer above a city may stop smoke and fumes from rising high into the atmosphere and flatten them into a thick layer hanging over the city. Two ways of preventing this are: (1) to build tall chimneys that help smoke to rise above a low inversion layer; and (2) to make sure that smoke is hot on leaving the chimney, as this helps it rise faster through the inversion.

and in Detroit, New York, and Chicago the annual deposit is more than 730 tons per square mile. One way of cutting down pollution is to build high chimneys and to make the exhaust as warm as possible so that the smoke plume has a better chance of rising above any inversion into the stronger winds and generally unstable air of higher levels. During stable atmospheric conditions in winter, large quantities of pollution are trapped at low level, and almost always the densest concentrations occur near the ground, though these usually fall off rapidly near the edge of a city or over an open space.

The pall of pollution that hangs above a city reduces radiation and sunshine near the ground by day, but, by radiating energy back, it helps to increase temperatures at night. The daytime screening effect is more pronounced than that at night, for absorption is strongest in the short wave-lengths. Very little ultra-violet radiation (from the sun) is able to penetrate a really dense pollution haze, and for this reason (among others) many city-dwellers suffered from rickets during the 18th and 19th centuries. Better diet and longer holidays as well as improved combustion methods have fortunately almost eliminated this disease in Europe. Even so, people who live in cities seldom have a sun tan. In December and January, central London receives less than 50 per cent, and suburban areas about three-quarters of the sunshine outside the city, even at rooftop level. On gloomy winter days as much as nine tenths of all radiation may be lost.

Suspended pollution particles also reduce visibility in three ways: by their own screening effect; by acting as condensation nuclei; and by discolouring the fog droplets to form a thick smog, aptly called a "pea-souper" by generations of Londoners. There is also an indirect as well as a direct connection between smoke and fog, for the same conditions encourage radiation fog as discourage the rapid dispersal of smoke. Fog (defined as a visibility of less than 1100 yards) occurs about twice as often in central London as in the surrounding countryside. But really dense fogs, with visibilities below 44 yards, last about four times longer in the suburbs than they do in the city centre. The interaction of temperature and pollution in cities produces an interesting cycle of visibility. Often fogs first form during the evening in the cool, humid air above the fields surrounding a city so that they encircle a relatively clear built-up area. During the night, temperatures fall, and the fog spreads over the central core until by dawn, it masks the whole city. Shortly after dawn, rising temperatures and increased turbulence probably disperse the rural fog, while in the city, where temperatures increase more slowly and winds in the streets are lighter, fog is more persistent. Town fogs also last longer because they absorb more solar radiation than cleaner rural fogs, so that the ground beneath is cooler and there is less overturning in the fog layer. Another reason for their persistence is that the acidity of the droplets may delay evaporation. So the pattern of early morning fog in a city is often the reverse of that during the evening. It consists of a blanket of fog that coincides roughly with the built-up area, but which is probably densest in the suburbs. Later in the day, the fog usually clears, even from the city streets.

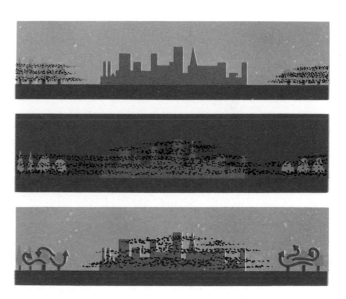

Top diagram, fog forms in rural areas during the day, but the town keeps clear because of its heat. Middle, at night the fog drifts into the city from the countryside. Bottom, the country areas soon clear next day, but fog lingers in the sheltered city.

The uneven surface of cities reduces mean wind-speeds but increases turbulence. As a result, city winds are generally lighter but more gusty than elsewhere, though these gust speeds may, during generally stable night-time conditions for instance, exceed those beyond the city. In a maze of streets between tall buildings, winds are very variable. There is a channelling of airflow along streets that run like man-made canyons parallel to the wind, and strong eddies where streets and winds intersect one another. On turning a corner, therefore, one can emerge from a gentle breeze into a full gale. Scraps of paper and dust scurry across a street, and rise up the lee face of buildings, while on the opposite side of the road, smoke is brought down to ground level. Cross-eddies and the jet-like funnelling of winds through gaps between or under buildings can often be avoided by careful planning. There is little or no remedy once buildings are up.

Up to a height of about 400 feet or so, most cities are enclosed in a mass of warm air known as a *heat island*. This island of air is heated in various ways, and its shape and intensity depend on the prevailing weather, particularly on cloud amounts and wind speed. At dawn, city temperatures are usually several degrees higher than those outside, but during the morning and early afternoon the city warms less quickly than the neighbouring fields and woods. This is due partly to the different heat capacities and conductivities of the fabric of a city and of vegetation-covered soils; partly to a haze hood, fog, or cloud above the built-up area; and partly to the mixing of warmer air near the ground with higher, cooler air by turbulence created by the city's uneven surface. So rural

An experiment to help town planners and architects. Models of a low and a high block of buildings are placed in a wind tunnel; the movement of winds around the blocks is made visible by smoke and carefully charted. Above, a photo of the experiment. Right, a drawing of the main air currents.

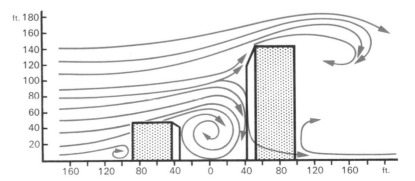

areas soon warm to temperatures that are the same or even higher than those in the city streets, in spite of evaporation cooling and the possibly higher albedo of woodland and farmland on the one hand and the heat of combustion released into the city air on the other. At night, some of these factors cause rural temperatures to fall more rapidly than those in the city, where stored solar energy is released from buildings, roads, and pavements, and where fires, furnaces, and engines help to warm the air. The enormous consumption of fuel in a large city may supply energy equal to one third of the sun's radiation falling on the city but its contribution to the heat island of many cities is probably small even where the air is stable. The heat islands of London, Manchester, and Washington, for example, are strongest just before dawn, when most fires are no longer burning, and in summer and autumn when they are rarely lit. On average, central London is warmer by 1.9°c around dawn and by 0.6°c in the afternoon, giving a mean deviation of 1.2°c. On calm, clear nights in late summer and early autumn, temperatures in central districts can be as much as 9°c higher than those above fields in the Green Belt.

The largest city does not necessarily have the largest heat island, since its intensity is closely related to the density of the buildings. Temperatures are high where buildings are close together and streets are narrow; they are low over parks and wide roads. Since large buildings are generally concentrated near the centre of the towns, it is here that the greatest deviations in temperature are generally found. Compact villages can also generate surprisingly large heat islands.

The extra warmth of cities is also indirectly responsible for other features that distinguish the climate of cities from that of their rural surroundings. In cities, relative humidities are reduced by an average of 5 to 10 per cent by higher temperatures and by the rapid draining of precipitation into sewers, although on some nights when the heat island is particularly intense, the reduction may be as much as 30 per cent. Absolute humidities on the other hand differ much less; in fact, on many nights they are higher in city streets

than above rural fields. There are two probable explanations: first, more moisture is lost from the air in rural areas by the formation of dew; and second, high humidities build up during the day in the warm air of city streets, and this air is trapped at night between buildings, and is thus prevented from mixing with the drier air above.

Contrasts in cloudiness and precipitation between town and country are more complex and less well understood, since it is hard to obtain comparable records from urban and rural stations, uncomplicated by topography and other such influences. However, it seems that many cities tend to be more cloudy, particularly in summer. This is probably because of increased turbulence over the built-up area. Isolated stationary clouds have also been reported above steam-cooling towers, factories, and blast furnaces.

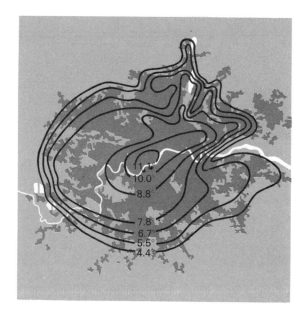

Temperatures are usually somewhat higher in a city than in the surrounding rural areas; this difference is more marked during the night. The chart above shows minimum-temperature isotherms (in degrees Centigrade) in London on the night of May 14, 1959. A deep anticyclone over Europe allowed the formation of an intense heat island over the city.

131

# 12　Past Climates

The most constant characteristic of world climates is their inconstancy. Many of the fluctuations are short-lived and apparently arbitrary; others reflect a fundamental shift in climate that may persist for hundreds or even thousands of years. We have only to look at the former climates—or *palaeoclimates*—of north-western Europe during the last million years of geological time (known as the Quaternary Period) to find examples of both major and minor changes.

During much of the Pleistocene Era (the earlier part of the Quaternary Period), which ended roughly 10,000 years ago, 8 million square miles of northern and central North America, Europe, and north-west Russia were covered in a sheet of ice, which in many places was more than 10,000 feet thick. (In the Southern Hemisphere, parts of south-east Australia, New Zealand, Patagonia, and southern Chile were also masked in ice.) But the Quaternary Ice Age was not continuous: there were breaks in the glacial conditions— some lasting several centuries, some many thousands of years—when the climate was often milder than it is today.

In Britain, the last ice age reached its final peak about 17,000 years ago, but by 8000 B.C. the country was almost clear of ice. As the ice sheets melted, the North Sea rose and broke through what is now the Straits of Dover, and Britain became an island. From 1000 to 500 B.C. the climate became far damper, with much cooler and cloudier summers. Later changes are of course mentioned in contemporary records; two of the most persistent were the warm, reasonably calm period

from A.D. 1000 to about 1250, and the "little ice age" between about 1550 and 1850. During these three centuries, alpine glaciers grew and arctic sea ice advanced to its most southerly position for 10,000 years.

From 1850 to 1940 there was another change in climate: a widespread warming with a tendency toward more maritime conditions on both sides of the Atlantic. The warming was greatest in the Arctic, where temperatures probably rose by as much as 2.8°C, while south of about 50°s there was little or no change at all. In many areas— Britain, for example—there were also increases in cloud and rain, although aridity increased in the sub-tropical deserts.

This remarkable warming of the climate had a considerable impact on plant and animal life, which advanced to higher latitudes and to higher levels. Marine life also spread as the seas became warmer. By 1938, sea ice in the Arctic Ocean had retreated farther north than ever before in historical times, and northern ports remained ice-free for much longer yearly periods. It was even predicted that if the improvement continued, the whole Arctic Ocean would become open by the end of the century. The yearly duration of sea ice around the coast of Iceland fell from 12 weeks in the 1880s to $1\frac{1}{2}$ weeks from the 1920s to 1940s. But in the 1950s, the duration rose again to about 4 weeks. This last figure is only one of many indications that the warming of the first forty or fifty years of the 20th century is over; the present trend is toward a harsher climate. In most areas, the reversal began in the late 1930s and

During the last two thousand years, there have
been many occasions when the Thames at London has
frozen over, allowing people to walk across the
ice in safety. In particularly cold winters, the
ice has been thick enough for fairs to be held
on it. This contemporary painting shows the fair
of 1684, seen from the south bank. In recent years
the climate has been warmer and embankments have
increased the speed of the river so that thick
ice has not formed.

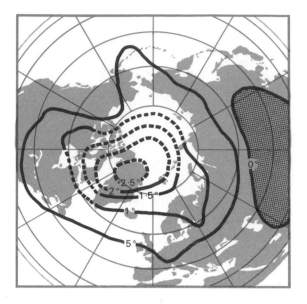

This map shows the increase in mean annual temperatures (in degrees Centigrade) in the area surrounding the North Pole from the period 1881–1938 to the period 1929–1938.

the 1940s, though in the eastern United States it seems to have delayed until the early 1950s. The American meteorologist J. M. Mitchell has shown that the world-wide average temperature has fallen by between 0.1 and 0.2°C since 1940, after having risen by 0.5°C between 1880 and 1940.

Clearly, such recent alterations in climate are easier to identify than are those of earlier times. But even without written records, there are still many other sources of information about the climates of prehistory. There is in particular a great deal of geological evidence; for example, boulder clays and frost-shattered layers are indications of glacial or near-glacial climates, while coal seams are the fossilized remains of plants that grew in warm, rainy areas. Similarly, by studying the actual formation of the land, we can locate the position of former snow-lines and so deduce the temperatures and precipitation of glacial periods.

The analysis of lake deposits can also yield valuable data. Traces of pollen and of invertebrates that had shells or hard skeletons indicate ancient environments similar to those in which the same organisms flourish today, while thin layers (or *varves*) of deposit in lakes have been used to study and date seasonal changes such as the annual summer melting of ice. In the sea similar information can be gained from deep bores through the sea-bed, and in the polar regions deep drilling of snow-fields reveals the precipitation from year to year. We have even been able to build up an idea of the chemical composition of past atmospheres from air trapped in snow that fell many hundreds of years ago.

Archaeology as well as geology has contributed to our knowledge of former climates. The discovery of ancient rock drawings of hunting scenes in the Sahara, for example, suggests that the area was not always entirely desert. Similar evidence is provided by petrified forests in lands where there is now very little vegetation, and by coffins penetrated by plant roots in parts of Greenland that today are permanently frozen. Such evidence of former climates is even more valuable if we can assign an accurate date to it. For this reason, the recently discovered technique of *carbon dating* is particularly important. All living matter, whether animal or vegetable, contains carbon that came originally from the photosynthesis of atmospheric carbon dioxide in plants. (Animals of course absorb carbon indirectly, by feeding on vegetable matter or on other animals that have lived on a plant diet.) Atmospheric carbon dioxide contains two types of carbon: the first is ordinary carbon with an atomic weight of 12; the second is a radio-active isotope of carbon known as Carbon-14, which is formed by the bombardment of atmospheric nitrogen by cosmic rays. When an organism dies, the small quantity of Carbon-14 that it has absorbed during its lifetime breaks down at a known speed. It is therefore possible to determine the age of an organism by measuring its Carbon-14 content. There is one limitation, however: any object over 30,000 years old contains so little Carbon-14 that its age cannot be determined by this method.

Below, a rock in Inverness, Scotland, scratched by the movement of a glacier during the Great Ice Age. As it moves along, the ice of a glacier picks up small pieces of rock from the ground, and the pressure of the ice above forces these against the rocks over which the glacier travels, making deep grooves (or striae) in the surface.

Above, a section cut through the earth during construction of a road near Southend in 1922, revealing many different layers of glacial deposits. (From surface: sandy loam, brown clay and loam, grey clay, white sand, and buff loam.) Each layer probably represents a rather different condition of the ice in response to changes of climate.

For climatic changes during historical times, there is a mass of written evidence. For example, records of temperature have been kept in many countries of north-west Europe and their overseas colonies since the late 18th century, though these were fragmentary until the mid- or late 19th century. There is also a great deal of indirect evidence, such as the dates of harvesting in the vine-growing areas of Europe since about 1400. Even so, it is dangerous to draw any definite conclusions from such information, since the events, like all weather changes, may have been caused by many different meteorological and non-meteorological factors.

These then are a few of the ways in which evidence of past change is assembled. Our next task is to attempt to explain how these changes came about. But before doing so, we must remember that climate is made up from so many different variables that the same explanation cannot be applied to, say, the change from Carboniferous to Permian times some 200 to 250 million years ago as to the general warming during the first 40 years of the present century. But all changes of climate, whether minor variations or major shifts, can ultimately be traced to alterations in the general circulation. And here we are immediately confronted with our lack of knowledge of the atmosphere—an ignorance that makes it extremely difficult to assess the changes necessary to explain past climates. We have seen that most seasonal climates vary enormously from year to year, although the individual factors causing these changes often seem to be unimportant. In other words, short-term variability may be a built-in characteristic of the atmospheric system rather than the result of any extra-terrestrial influence such as sun-spots and so on. To be attributable to such an outside cause, a trend must be established over at least 10 years, so that it is statistically improbable that it could have occurred by pure chance.

In any discussion of the general circulation, we must remember that the oceans and ice-caps, because they are slow to warm compared with land masses, act rather like gigantic fly-wheels in the mechanism of the atmospheric heat engine. It has been suggested that temperature fluctuations persist for hundreds of years in the deep oceans, and for tens of thousands of years in the major ice-caps. And quite small alterations can precipitate a series of atmospheric events that may enhance their overall effect. For example, an ice-cap, once formed, is likely to persist because its high albedo reduces the amount of melting by the sun. It also tends to grow because the huge anticyclone that rests above it steers local storms around its edge where they deposit yet more snow.

Anything that disturbs the earth's energy balance can influence climate. One cause of disturbance can be changes in the configuration of land masses. For instance, the formation of the Rocky Mountains, the Andes, and the Himalayas about 30 million years ago must have caused profound alterations in climate. Even more radical was the major adjustment of the distribution of land masses and oceans that probably took place some

time in the last 200 million years. This idea of *continental drift* was first put forward by the German meteorologist Alfred Wegener about 50 years ago. It proposes that in Carboniferous times there were two main land masses, one called Pangaea, which included Europe, most of Asia, Greenland, and North America, and the other Gondwanaland, comprising Africa, India, Madagascar, South America, Australia, and Antarctica. A series of splits then occurred: North America moved away from Pangaea, while the components of Gondwanaland also broke away and drifted into their present positions.

Wegener originally developed his theory in order to explain the fact that certain areas now in the tropics were once covered with ice, as is shown by the scars of glaciation and other evidence. He also sought the reason for the presence of coal seams, evidence of a once warm, wet climate and profuse vegetation, in the now permanently frozen Antarctic continent. But until recently, this theory of continental drift was rejected by many scientists, mainly because no one could find a convincing explanation for the tremendous forces required to fracture the original land masses and to shift them over distances of thousands of miles. Today, however, many geologists believe that these forces were produced by convection currents set up in the earth's mantle by heat generated by radio-active break-down. And as some of these currents moved upward toward the earth's crust and then flowed apart, sufficiently powerful tensional stresses were built up to split the continental masses and to set them drifting on the crustal surface. Such a theory accounts for two major climatic phases: the pre-Carboniferous distribution of land masses and hence climates, and later the drift of continents into approximately the arrangement we see today, with yet another totally different set of climatic conditions.

Another way in which the earth's energy balance can be altered and climates can be modified is by events outside the earth—for instance, by a variation in solar radiation. The results of such a variation were set out in an ingenious theory put forward by the British meteorologist Sir George Simpson to

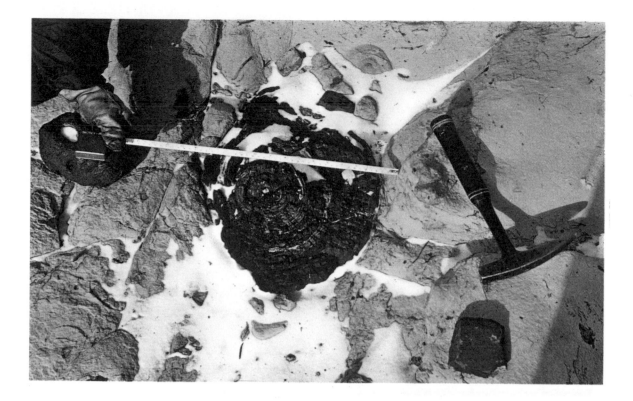

Above, this fossilized tree trunk was found during geological excavations in the Antarctic. It proves that the climate of this region was once able to support trees, although today no vegetation except lichens can survive. Right, a cross-section of a fossilized tree trunk found in Washington State, U.S.A.; this area can now support only sage-bush scrub, indicating that a change of climate has taken place in the region.

explain the Quaternary Ice Age. According to Simpson, this great glaciation was not caused by a decrease in solar radiation, as one might expect. Such a decrease, he argued, would lower the temperature of the earth's surface, particularly in the tropics. This drop in mean temperature would have two effects: it would slacken the poleward temperature gradient and decrease evaporation in tropical areas, which would in turn reduce the overall water-vapour content of the atmosphere. Inevitably, the general circulation would be weakened, and storminess and precipitation in all latitudes would be reduced—conditions known to be the opposite of those in glacial periods.

Simpson therefore concluded that the changes necessary to produce glaciation would be brought about by a small *increase* in solar energy, which would intensify cloud and precipitation, particularly in upland areas near the poles. The increased cloudiness would also reduce temperatures in high latitudes, and ice-caps would then spread out. This theory has provoked some criticism, although the basic premises seem perfectly sound. Several modifications have also been advanced, one of which suggests that glacial periods are encouraged by a brief increase in solar radiation occurring within a longer period of generally low radiation. The earth, especially the oceans in high latitudes, would then have been cooled in preparation for the growth of an ice-sheet.

While admitting that Simpson's theory is possibly correct, it is only fair to mention that the Yugoslav physicist A. Milankovitch has reached the opposite conclusion by relating the ups and downs of the Quaternary glaciations with estimates of the amount of solar energy actually reaching the earth over the past million years. Milankovitch's calculations, based on the shape of the earth's orbit at different periods and on the angle of the earth's axis to the orbit, showed that the periods of greatest glaciation tally exactly with those periods when the earth received the least amount of heat from the sun. However, there is no way of knowing what fluctuations in the sun's heat production may have occurred over such a long period.

For recent times, no conclusive evidence of any variations in solar radiation has been recorded—though this does not necessarily imply that they have never happened. The sun is known to go through both long- and short-term cycles of change in state. These changes are confirmed by observations of sunspots, whose number and area have been recorded since the mid-18th century. Sunspots seem to occur in cycles that have an average length of 11.2 years. Many attempts have been made to relate this period to hydrographic and climatic cycles, with interesting but far from conclusive results. Recent measurements of the solar constant indicate a variation of no more than 0.2 per cent.

By increasing the amount of carbon dioxide in the air, vast forest fires may possibly have a very slight influence on climate. Also, deforestation may lead to an increase of desert area, which, similarly, may slightly affect climate.

Such a change would probably produce an atmospheric temperature fluctuation of less than 0.1°c—almost ten times less than that observed during recent times. So, apart from indirect climatic data, there is little evidence to either prove or disprove that larger, or at any rate larger than present, solar variations have occurred over long periods of time.

Recent evidence seems to show that if the answer to climatic change does in fact lie in variations of solar radiation, then it is to be found outside the 0.3 to 3 micron band—the wave-length that dominates the radiation reaching the levels at which most measurements are made. In other words, we need to concentrate on ultra-violet and corpuscular radiation. It is well known that during solar flares, far more X rays, high-energy photons, and particles are emitted in the ultra-violet range of the solar spectrum. Of this short-wave radiation, wave-lengths of less than 0.1 microns are mainly absorbed in the ionosphere at heights of about 62 miles, while slightly longer-wave radiation in the 0.1 to 0.3 micron band is absorbed lower down, mainly by ozone at heights of about 35 miles. It has been suggested that variations in the amount absorbed at these levels might lead to long-term changes of temperature and pressure, which could in turn affect the troposphere below. So it has been argued that a variation of ultra-violet radiation emitted by the sun can modify the pattern of circulation almost without affecting the solar constant or the total amount of energy used to heat the lower atmosphere. Even so, evidence from the solar flare of November 12, 1960—perhaps the most intense of the last 30 years—provided little support for this theory; nor was it confirmed by an examination of the behaviour of the ozone layer during solar flares.

So far, then, attempts to establish variations in solar radiation as a factor of climatic change are inconclusive. But before leaving the subject, we should perhaps mention one fascinating new theory, which proposes that the whole solar system sometimes passes through areas of interstellar dust. Between one half and two thirds of the total mass of matter in our galaxy is thinly and rather unevenly distributed in space. This interstellar matter often collects into clouds that are about 100 times denser than the distribution of mass in surrounding space. No one is sure what would happen if the solar system should pass through such a cloud, but many agree that the sun would attract and collect some of this dust; this would be added to the sun's own stockpile of heat-producing material and would increase the output of solar energy by something like one per cent, most of which would lie in the ultra-violet range.

Radiational energy receipt and loss, either regionally or on a global scale (and probably both), are also affected by changes in the composition of the atmosphere. Some of these changes can be directly attributed to human activity; whether they in fact have any appreciable effect on climate is less clear. As we saw in Chapter 2, the average concentration of carbon dioxide in the air has risen from 290 to 330 parts per million since the beginning of the century. Most of the increase can be ascribed to the burning of fuels, but

some of it may be due to other causes: these include a rise in bacterial activity in the soil of the far larger areas now under cultivation, the decomposition of greater amounts of animal and vegetable matter, and more bush fires. Carbon dioxide, as we saw in Chapter 3, intercepts long-wave radiation, and so any increase must tend to raise temperatures in the lower troposphere. In fact, it has now been calculated that the observed annual rise in temperatures corresponds very closely to that attributed to the increase of carbon dioxide during this century—$0.01°c$ per year. Unfortunately, the correspondence is too exact, since it leaves no margin for the effect of other observed changes such as pollution. It also fails to explain the almost constant level of temperature south of 50°s—areas where the carbon dioxide content of the air is much the same as elsewhere. Finally, there have been wide and inconsistent fluctuations in temperatures ever since the early years of the Industrial Revolution, culminating, as we saw earlier in this chapter, in a period of falling temperatures since about 1940. This variation seems hard to reconcile with the fairly consistent increase in the output of carbon dioxide, though it has been suggested that concentrations have probably fluctuated because of a complex cycle of exchange between the atmosphere and the oceans.

There has also been an enormous increase in the amount of smoke expelled into the atmosphere during the last two centuries owing to the widespread burning of coal in domestic grates and industrial boilers. Dust specks scatter and absorb solar radiation, but this absorption also results in an increase in the downward radiation of infra-red wavelengths, which would probably compensate for the loss of direct solar radiation reaching the ground. The overall effect of smoke particles is, however, negligible, since even the smallest rarely remain in the atmosphere for more than a few days. Dust from volcanoes behaves rather differently. The larger particles fall to the ground fairly quickly, but vast quantities of the smaller ones are lifted into the stratosphere where they remain for a period lasting from one year in high latitudes to four in low latitudes. Obviously, this suspended dust greatly increases the scattering of solar radiation, but other changes will also follow. Evaporation and convection—and so cloudiness—usually tend to decrease, so that reduced infra-red losses do to some extent compensate for any loss in direct solar radiation. Even so, it seems that there would still be a small net reduction in energy. A great advantage of this theory is that the effect of volcanic dust can be measured directly. After the Krakatoa explosion of 1883, dust remained in the atmosphere for at least three years, and the amount of solar radiation recorded at Montpellier, France, fell by 10 per cent during this time. There are similar records for other notable volcanic eruptions.

These then are some of the explanations put forward to explain climatic change. Some are more convincing than others, but none is completely satisfactory. For this reason, many palaeoclimatologists have tried to identify underlying cycles or rhythms in the records of past climates. Such periodicity analysis has often been unhelpful, not because there were no cycles to be found, but because there were too many.

Understandably, theories of climatic cycles receive little credit today. A far better approach is by way of past circulation studies. For example, mild maritime conditions in Europe and North America—such as those prevailing at the end of the last century—are generally associated with a dominantly west-east (or zonal) circulation, with frequent depressions travelling unusually far north. But it has been estimated that since the 1920s the average track of depressions has moved 2 to 5° farther south—that is, between 120 and 300 miles—a shift that has resulted in cooler, wetter summers and harsher winter weather with more frequent northerly and north-easterly winds. At the same time, there have been corresponding changes in the patterns of circulation in the middle and upper troposphere. The recent tendency has been for an intense circulation through two to four long waves in the upper westerlies to be replaced by a weaker north-south and south-north (or meridional) circulation through shorter-wavelength waves of greater amplitude.

Above, an engraving, based on a very early photo, of the eruption in August 1883 of Krakatoa, between Java and Sumatra. The explosion destroyed two thirds of the island, and was heard 3000 miles away. It threw vast quantities of dust into the atmosphere where it remained for over three years. Suspended volcanic dust like this increases the scattering of solar radiation and generally reduces the amount of solar radiation reaching the earth. The possible climatic changes resulting from such a reduction are discussed in the text. The dust from the Krakatoa explosion caused extraordinary sunsets in many places. The paintings right, by a contemporary artist, show three stages of the sunset at Chelsea, London, on November 26, 1883.

# 13 Climate and Man

Man's success as an inhabitant of this planet has largely depended on his ability not merely to use but to transform his environment. Yet the atmosphere, one of the most important and complex features of his world, has for the most part remained beyond his control. Even so, he has never entirely put aside his dreams of consciously and directly influencing at least local climatic conditions. In this chapter, we shall be looking at both sides of the interrelationship between climate and man: how climate affects man; and how man has tried to modify climate.

Let us start with the way in which climate influences man most directly—by affecting his physical condition. We are all aware that certain days and certain places are more invigorating, more stimulating to mental and physical activity, than others. One explanation, which has been confirmed by recent research, was first put forward by the American geographer Ellsworth Huntington (1876–1947). According to this theory, a monotonous climate is far more enervating than changeable weather; temperate cyclonic climates are therefore considered particularly stimulating, whereas tropical and continental interior climates are thought to encourage physical and mental lethargy. It seems that physical efficiency is highest when temperatures are around 18°C and relative humidities between 75 and 80 per cent, though ideally both should fluctuate, as should pressure.

When considering the impact of climate on man, the key factor is the capacity of the body to regulate its temperature. When relative humidities are high and temperatures are above 20°C, natural cooling by radiation and evaporation from the skin and lungs is insufficient to stop the body temperature rising above normal. Vigorous exercise is then followed by discomfort and a sense of oppression, and in extreme cases by heatstroke. Air movement obviously helps to increase the body's cooling power, even when temperatures and humidities are still high. For this reason, fans—or, better still, air-conditioning—are almost essential for indoor comfort in the humid tropics.

Until recently, we had little practical experience of the effects of cold climates on the human body. Today we know a good deal. In temperate climates, more than half the heat lost from the body is lost by radiation. In the tropics, radiation losses are small, but in polar regions they may rise to more than two thirds when the air is calm. In cold, windy conditions, a great deal of heat is also lost by conduction and convection, but much less by evaporation since the body perspires very little at low temperatures. Changes in relative humidity therefore have far less bearing on body comfort in cold climates than they do in the tropics. Actual physical injury from intense cold occurs when the body loses more heat than it can produce; this takes place when the body is immersed in cold water or exposed to cold air. Water has a cooling power 23 times greater than that of air, and at near-freezing temperatures may kill a man in less than 15 minutes.

Many human ailments and diseases have a clear association with certain types of weather. In some cases, there is a direct physical

Top, this scene in a Mexican street is evocative of the atmosphere at midday in a hot, enervating climate. According to Huntington, the people of regions with such a climate tend to be passive and apathetic.

Bottom, the variable temperate cyclonic climate is extremely stimulating. Such a climate produces a restless, energetic population; even their leisure time tends to be vigorous and organized.

143

A Spanish street showing how the traditional architecture developed to suit the local climate. Thick white-painted walls and small windows help keep houses cool by day and warm by night.

relationship between climate and illness—for example, the close connection between bronchitis and polluted, foggy atmospheres. Often the association is less direct but still unmistakable. For example, certain climates obviously foster the development and spread of particular viruses, organisms, and disease-carrying insects such as the malarial mosquito and the tsetse fly (which causes sleeping sickness). Recent research has also shown that even certain old wives' tales about the weather may sometimes be based on fact. Some experts believe that old wounds, corns, and aching joints can, as claimed by many sufferers, forecast the weather (especially rain) by reacting to electrical disturbances that herald the arrival of fronts.

Clearly, climate also plays an important part in the development and spread of animal and plant pests and diseases. To take a well-known example: research shows that potato blight is extremely likely to develop about two weeks after a 48-hour period during which minimum air temperatures (taken at four feet) have never fallen below 10°C, and the relative humidity has remained above 75 per cent. Such conditions often occur in the summer during prolonged wet, overcast weather brought by slow-moving depressions. Some diseases that are not local in origin depend even more directly on the weather: germs, spores, and insect pests can be carried by the wind over hundreds of miles. The spores of black rust, a disease of cereal crops, can reach southern England from as far away as North Africa, and there are even reports of locusts surviving this journey in suitable wind conditions.

These are just a tiny sample of the effects of climate on man and his environment. We now turn to some of the ways in which man has sought to modify these effects. Many of these involve isolating and insulating parcels of the atmosphere inside buildings to create a comfortable living or working climate. In hot countries, buildings should be sited and insulated to cut down the effect of strong sunlight and to encourage ventilation by local breezes. In Israel, for example, care has been taken in planning coastal cities to leave gaps between the buildings along the sea front to allow the cooling sea breeze to move inland. The relative temperatures inside and outside buildings depend of course upon the albedo and conductivity of the walls and roof. Thick, white-painted walls with small windows keep North African buildings cool by day and warm by night. Cavity walls plus double-glazed windows also help to insulate a house from cold outside temperatures. A house's aspect is also important: on a cold winter day in the Northern Hemisphere, five times more artificial heat may be needed to warm a room facing north than to heat one facing south.

The same principle of insulation applies to the successful growing of crops indoors. Even in an unheated greenhouse, the temperature seldom falls below $-2°C$. A low glass frame provides better protection; it is about twice as effective as a greenhouse. And if the frames are sprinkled with water so that a coating of ice (a poor conductor of heat) forms on the glass, the air inside the frame can be kept up to $10°C$ warmer than that outside on a cold winter night. However, in many glasshouses the risk of frost is entirely eliminated by artificial heating and by water sprinklers.

Even outdoors, crops can often be protected from damage by frost—perhaps the most serious weather threat confronting farmers and fruit growers in the mid-latitudes. Since prevention is always better than any cure, the choice of a frost-free site is obviously better than any remedial measure. If this is impossible, suspending canopies above the crop to reduce night-time radiational heat losses from the ground is one of the simplest methods of protection. Smoke screens are more spectacular, but they are less easily controlled and may cover the leaves in harmful soot and provoke complaints from neighbouring farmers. Another popular way of combating frost is to warm the air by burning some kind of cheap fuel—usually heavy oil—in what are known as smudge pots, which can raise temperatures by as much as

$5°C$ if the warmed air is kept near the ground by a low inversion of temperature. Large-vaned rotating fans placed just above the crowned layer of fruit trees can also help to prevent frost by creating turbulence that stops cold air settling near the ground.

Wind as well as frost can cause serious damage to crops. For many hundreds of years, the inhabitants of such areas as the broad valley of the river Rhône in France have planted hedges and trees or built walls to protect their farmland from the ravages of drying and destructive winds. Obviously the amount of protection depends on the type of shelter. In the lee of a solid wall, there is a dramatic drop in wind speed, though once over on the other side, wind picks up rapidly and recovers its original speed at a distance equal to only about five times the height of the wall. But this type of solid shelter also causes turbulence and gusty winds, so that it is something of a mixed blessing except in a small walled garden. A loosely knit fence or hedge gives less protection, but without eddying and harmful gusts. The best type of shelter is therefore a barrier such as a stand of trees that is denser at the top than at the bottom, so that air filtering through the bottom prevents eddies from forming in its lee. Directly in the lee of such a shelter-belt, winds are reduced to less than a quarter of their unobstructed speed. Above the shelter-belt and around its ends, however, winds increase in speed, so it is important not to leave any gaps in the barrier. Obviously, such barriers can shelter only quite a small area, but a series of parallel screens, placed at right angles to the dominant or most damaging winds at intervals of about 30 times their height, can protect several square miles.

Hail is a major threat to crops and buildings in many countries. Practical attempts to prevent hail damage have a long (and sometimes faintly comical) history, notably in the vine-growing areas of southern and central Europe. All of these were based, as we now know, on the use of pressure waves. As early as the eighth century in Europe, church bells were rung when storms threatened. From the 18th to the early 19th centuries, guns were commonly fired into thunder-clouds. Today

The Mistral is a cold, northerly local wind that threatens crops when it blows down the Rhône valley. Hedges, walls, and lines of trees have been arranged across the valley, to shelter the crops from the wind. The photograph above shows a typical section of the valley, and the diagram below shows the effect of the barriers.

The wind is forced to rise over and around the barrier, and its velocity is considerably reduced in its lee. It gradually picks up speed again, but can be further slowed by another barrier. Thus, careful spacing of the wind-breaks permits good crops to be raised successfully where they would otherwise be killed by the strong, cold wind.

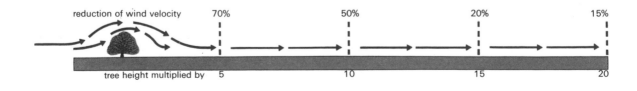

reduction of wind velocity   70%   50%   20%   15%

tree height multiplied by   5   10   15   20

hail cannons have been superseded by rockets, and in France, Italy, Russia, America, South Africa, Kenya and elsewhere, large areas of farmland are covered by a network of rocket-firing stations, usually manned by the farmers themselves. The rockets explode at altitudes of 4500 to 6000 feet; the resulting pressure waves are thought by some to shatter the naturally compact stones, turning damaging hard hail into soft hail or rain.

Experiments to improve the efficiency of rockets and, perhaps even more important, research into the physics of hail formation are being carried out in Switzerland, Russia, the United States, and several other countries. Attempts have been made to prevent both hail and lightning by over-seeding potential thunder-clouds and so giving the supercooled droplets a chance to freeze before they can grow by sweeping (p. 90). It is doubtful, however, whether it will ever be feasible to collect together the prodigious amounts of freezing nuclei that would be needed.

Like hail protection, ways of making rain —a crucial element in any climate—involve actual interference with the atmosphere. In many parts of the world, there is not enough rain to water the crops, or it falls too early or too late. Where lack of rain is a constant hazard, rain-making rituals have frequently developed as an important part of the religious life of a community. But rain-making, whether it relies on magic or on science, is attempted only when rain seems likely— that is, at the beginning of or during the rainy season. Close observation of clouds can help a witch-doctor to become a reasonable weather forecaster, even though he may represent his science as a magic art. Early experiments to increase rainfall involved such varying techniques as the lighting of bush fires, the firing of cannons (sometimes using shells filled with liquid carbon dioxide), the spraying of dust from balloons, and the production of electric charges by kites.

Truly scientific rain-making began with an experiment over Prague in September 1942. This was carried out by the distinguished German cloud physicist Walter Findeisen, who used a Heinkel aeroplane to spread silica dust into a small, supercooled cloud with a temperature of $-8.5°C$, some $10°$ warmer than the temperature at which ice crystals normally form on freezing nuclei in the atmosphere. Soon after the seeding, the cloud changed into a trail of falling snow crystals. This experiment was the first recorded example of a deliberate and successful attempt to modify a natural cloud, and it opened up a new, exciting vista of rain-making and other forms of weather control. But it was not until 1946 that a method was found of manufacturing freezing nuclei in sufficient quantities to convert the supercooled crests of clouds into ice crystals and so set the Bergeron rain-forming process in motion (p. 90). This happened when the American meteorologist Vincent Shaefer discovered that a fragment of solid carbon dioxide no larger than a pin's head would produce hundreds of millions of tiny ice crystals when dropped into a laboratory cold chamber containing a supercooled cloud. (Solid carbon dioxide—generally known as dry ice—has a surface temperature of about $-70°C$.) The first outdoor trials took place in November 1946, when a supercooled cloud was successfully converted by Shaefer's method into a dense trail of snow crystals. Soon afterward, Bernard Vonnegut, working in the same laboratories as Shaefer, made another important discovery about cloud seeding. He found that minute crystals of silver iodide, which has a crystal structure remarkably similar to that of ice, act as ice-forming nuclei at temperatures below $-5°C$.

Successful early trials with dry ice and silver iodide as artificial ice nuclei were quickly followed by the formation of rain-making companies by hopeful speculators. Few of these survive today, since rain-making was found to be far less simple than was first supposed. Only clouds of a certain form and size were found to respond to seeding; further, the dry-ice process was expensive, since it involved the use of aircraft. In the case of silver iodide, which is dispelled as smoke from ground generators, the particles often failed to reach the supercooled layers of the cloud; also strong sunlight breaks down the particles in the same way that silver bromide is affected in a camera film. For these reasons, seeding directly into the cloud by aircraft, at first sight more expensive, may in the long run be more economical. In fact, Australian scientists have now turned over almost entirely to airborne silver iodide experiments, more than two thirds of which seem to have had successful results. In 1965, wide-spread forest fires in New South Wales were partly quenched by rain from seeded clouds.

Today scientists in many countries are engaged on programmes of basic research, laboratory experiments, and field investigation into artificial rain-making. Most are based on the established fact that many supercooled-droplet clouds can be transformed into ice clouds. In Australia and Japan, however, a series of field investigations has been carried out that attempted to simulate and stimulate the formation of rain by the collision and coalescence of variously sized droplets in "warm" clouds. In the earliest of these experiments, cumulus clouds were sprayed from above with large water droplets, but without any noticeable result. This is not surprising, since a simple calculation of the rates of droplet growth shows that this technique is likely to succeed only when the cloud is very deep—in other words, when rain is likely to fall without encouragement. A more promising method is to introduce small droplets into the base of actively growing clouds, since droplets grow during both their upward and downward journeys through the clouds. During a series of recent experiments in Australia, rain fell from a number of clouds that had been seeded in this way. Another comparatively new technique uses dry salt crystals as nuclei. This last method imitates a natural process that occurs over the oceans, where there is roughly a sackful of salt nuclei in an average shower cloud.

So far we have mentioned a few of the most recent ideas and methods for changing or combating local weather conditions. But some scientists are already considering possible ways of modifying climates on a regional, continental, or even global scale. For example, a radical change in climate could be brought about by altering the amount of radiation absorbed by the earth's surface by adding to the atmosphere's quota of carbon dioxide, water vapour, and dust or by altering the earth's albedo. Already, in parts of Russia and Canada, snow is encouraged to melt earlier in spring by covering it with coal-dust, which absorbs more solar radiation. Some method based on the same principle has also been suggested for melting the margins of polar ice-caps. But such large-scale schemes arouse more alarm than enthusiasm, since we have little idea of their long-term effect. As we saw in Chapter 12, for example, it is still uncertain whether more radiation and a warmer atmosphere would increase the present rainfall and snow cover.

Today weather modification is the target of a lot of theoretical and experimental

This engraving shows a machine, patented by Daniel Ruggles of Virginia in 1880, in the hope of producing rain. An explosive charge was sent up by balloon into a suitable cloud, and then exploded from the ground electrically. It is unlikely to have worked.

148

research in more than 20 countries. But as more data accumulates, man is beginning to realize that the resources of energy at his disposal for modifying the weather are negligible compared with the forces at work in the atmosphere. About 10 million tons of air are set in motion to produce a fairly small shower cloud, which then goes on to absorb air into its base at a rate of about 100 million cubic feet a second. Often there are several million such clouds in a temperate depression. The total kinetic, thermal, and electrical energy of one thunder-storm is roughly equivalent to the energy released in a megaton-bomb explosion, while one single tropical hurricane may generate many thousands of times more energy than the most powerful hydrogen bomb yet produced. It is therefore extremely unlikely that man will ever be able to manipulate the weather to any great extent, though he will probably improve his present modest control of local conditions. To put it another way, man can conquer the weather only by strategy, not by superior force. In many cases quite small doses of energy released at well-chosen moments can modify local weather considerably. The chief danger is that major man-made changes in weather may have global repercussions. For this reason, there must be not only a greater degree of international co-operation but a deeper understanding of the basic workings of the atmosphere before man is allowed to tamper with the weather on any scale.

A more modern method of producing rain by artificial means is cloud seeding. In one technique, silver iodide crystals are released in a cloud to allow its supercooled droplets to freeze. In the cloud shown below the ice crystals have fallen out and left a hole.

# Index

Page numbers in *italics* refer to illustrations or captions; references combining two or more pages (e.g., 58-9) imply both text and captions

Adiabatic rate, *see* Lapse rate
Advection fog, 78, 114
Aerodynamic calculation, 67, 68
Aeroplane: used for meteorological recordings, 18, *29*; used for rain-making, 147
Aerosols, 79;
    in urban air, 128
Air: balance of forces on, 115; currents, 58-9, 62, 74, 79, 114, *130*; data about, 20, 22-3; dry, 72, 74, 92; "heat island" 130-1; humidity of, measured, 17, 68, 79; inversion of, 104, 118, *128*; moisture in, 70, 72, 74, 78-80, 123; movement of, 42, *43*, 44, 46, *46*, 47, 49-52, *58*, 67-8, 70, 72, 74, *74*, 75, 78-9, 84, 88, 110-1, 118, 120, 124, *125*, 142, 149; pollution of, 126; pressure of, 20, *49*, 111; saturated, 67, 72, 74, 79-80, 83; storms in, 92, 99; super-saturated, 79-80; temperature, 16, 26-7, 74-5, 87, 116, *119*, *120;* tropospheric, 115; warm, *17*, *70*, *72*, *73*, 115;
    *see also* Aerosols, Atmosphere, Lapse rate, Precipitation, Thermals
Airflow, in hailstorm, *89*
Air masses, 100-6, *107*;
    labelling of, 100; map of sources, *102*;
    *types:* arctic maritime, *102*, 103; polar continental, *102*, 103-4, 106; polar maritime, *102*, 103, 106; returning polar maritime, *102*, 103, 106; tropical continental, *102*, 104; tropical maritime, *102*, 103-4, 106
Aitken, nuclei, 80
Albedo, 38;
    of buildings, 145; of clouds, 38; of common surfaces, 38, 122; of earth, 38, 148; of forests, 122; of ice cap, 136; of rural areas, 131; of snow, 38, *122*
Aldehydes, 128
Altostratus, 76
American Weather Bureau, computer readings of, *60*
Anabatic wind, 120-1
Ana-front, *104*, 105-6
Angular momentum, of atmosphere, 59
Anticyclone, 46, 50, 52-3, 55, 88, 103-4, 114-5;
    and Ferrel westerlies, 58; and ice-caps, 136; and London heat island, *131*; balance of forces in, *51*; blocking highs, 114-5; influence on winds, 60;

sub-tropical, 54, 59, 114; warm, 115
Anvil cirrus, 83
Arctic front, 104-5
Aristotle, 12
Atmosphere, 10, 15, 18, 20-6, 27, 30-2, 35-42, 44-6, 59, 62, 66-7, 74-6, 78, 80-1, 136;
    air masses in, 100, 103-4; angular momentum of, 59; balloon, studied from, *19*; behaviour of, 22, 62, 136; chart of, *28*, *29*; circulation of, 60, *60*; composition of, 22-4, 26, 30-1, 58; energy in, 58; evaporation in, 62, 66-8, 75-6; ionization of, 58; man's control of, 142, 149; moisture in, 59, 66-8, 70; in motion, 42, 50, 53, 58-9; neutral stability of, 74; radiation at fringes of, 35-6, 40; reaction to radiation, 39; stable, 74; supply of solar energy to, 35-6, 37, 39; thermal structure of, *28*; in towns, 126, 128-31; unstable, 74; upper atmosphere, *19*, 30, *31*, 39, 100; vertical motion of air in, 74-6; warming of, 40-1, 47; water vapour in, 58;
    *see also* Air, Climate, Cloud, Condensation, Evaporation, Gases, Hydrological cycle, Precipitation, Pressure
Atmospheric effect, 39
Atomic oxygen, 22-3, 26
Atoms, 23
Aurora, *29*

Ball lightning, 95
Balloons: for rain-making, 147; for recording weather, 18, *19*, 22, 27, 30, *30*; sounding balloons, 19;
    *see also* Radiosonde
Barometer, 15;
    diagonal, *15*; invention of, 15; mercury, *12*
Bergeron, Tor, 88, 90, 106; rain-forming process of, 147
Bible, The, weather in, 12
Bjerknes, J., 104, 106
Bjerknes, V., 104
Blizzard, *120*
Blocking highs, 114-5
Bork, Léon Teisserenc de, 26-7
Bowen, E. G., 81, 90
Brooks, C. E. P., 95

Calorie, 35, 44
Carbon: carbon-14 dating, 134; in fuels, 26
Carbon dioxide, 19, 24-6, 36, 40, 134, 139, 147-8
Carbon monoxide, 128
Cell: Ferrel, 48, 50; Hadley, *46*, 47, *47*, 48, 50, 59; mid-latitude = Ferrel; polar, *46*, 48; reverse, 47-8
Central Forecasting Office, Bracknell, 100
Centrifugal force, 52
Charts, weather, *see* Weather charts;

*see also* Climate, Weather
Chinook, 120, *122*
Cirrostratus cloud, 62, 99
Cirrus cloud, 62, *64*, 76, *82*, 99
Climate, 10, 100;
    effect on animal and plant pests, 144; effect on man, 10, 142, *143*, 144-5, 147-9; historical survey of, 132, 134; local, 115, 116, *117*, 118, 120-4; past, 132, 134-7, 138-40; theories of, *10*; town, 126, *127*, 128, *128*, 129-31, *133*, *144*;
    *see also* Microclimate, Palaeoclimate
Climatic cycle, 138
Climatic tree-line, 116
Cloud, 23, 24, 36, 38, 41, 62, 66, 75-6, 78-80, 83-4, 88, 90, 92, *92*, 94, 104, 108-9, 120, 124, 131-2, 138, 148-9;
    as clue to atmosphere, 24, *29*, 62; classification of, 62; cover, effect of, 38; density, 88; diagram of, *64*; formation, *77*; in tornado, 97; layered, 75-6, 80, 84, 86, *90*; patterns, recorded, 19; precipitation in, 88, 90; rain-making, modified for, 147; seeding, 147-8, *149*; thunderstorms in, 94, 99;
    *types:* altrostratus, 76; anvil cirrus, 83; atomic bomb, 75; cirrostratus, 62, 99; cirrus, 62, *64*, 76, *82*, 99; cumuliform, 62, 66; cumulonimbus, *65*, 66, 75-6, *84*, 87, *87*, 90, 108, 112; cumulonimbus anvil, 112; cumulus, 62, *64*, 70, 75-6, 80, 83-4, 88, 90, *90*, *102*, 108, *115*, 124, *125*; Foehn-wall, 120; mother-of-pearl, 23, 79; nimbostratus, 76; noctilucent, 23, *29*; orographic, *73*; radiation fog, 75, *75*, 78, 114, 129; stratiform, 62, 75, *105*; stratocumulus, 38, 103, 108; stratus, 62, *64*, *75*, 76, 90, 104, 108, 114; thunder, 94, 147;
    *see also* Albedo, Atmosphere, Lightning
Cloud droplets, *see* Droplets
Coal, 26, 126, 128;
    dust, 148
Cold front, 105-6, 108
Condensation, 70, *70*, 72, 75, 76, 79-83, 88, 120
Condensation nuclei, *79*
Continental drift, *136*, *137*
Convection: and body's heat loss, 142; currents, 40-1, 81, 136; diagram of convective system of winds, *50*; movements of, 78
Coriolis force, 51, *51*, 124
Corpuscular radiation, 138
Crystals: ice, 82, *82*, 83-4, 86, 88, *89*, 90, *90*, *91*, 99, 147, *149*; shape of, 83; snow, 81-2, 83, *87*, *91*, 147
Cumuliform, cumulonimbus, and cumulus cloud, *see* Cloud
Cut-off depression, 115
Cyclones: balance of forces in, *51*; development of, 106, *107*;

# Acknowledgments

Key to picture position: (T) top, (C) centre, (B) bottom, and combinations, for example (TL) top left, or (CR) centre right

Endpapers by permission of the Controller of Her Britannic Majesty's Stationery Office; British Crown Copyright
6   Camera Press Ltd./photo Penelope Reed
11   National Maritime Museum, Greenwich
13   (C & R) Science Museum, London; British Crown Copyright
14   Science Museum, London/ photo Michael Holford
15   Science Museum, London; British Crown Copyright
16   *Phil. Trans.* vol. 16, 1686/ photo Michael Holford
17   (T) Photo Chris Ridley/by permission of the Director-General of the Meteorological Office
    (B) Science Museum, London/photo Michael Holford
18   By permission of the Controller of Her Britannic Majesty's Stationery Office; British Crown Copyright
19   (T) Mansell Collection
    (B) Diagram based on Rolf Engel
21   *Life* © Time Inc.
23   (B) The Press Association Limited
27   (T) Guide Bridge Rubber Co. Ltd. (Phillips group of companies)/photo Photoflex
    (B) Science Museum, London/photo Michael Holford
29   (R1 & 2) From the Diana Wyllie set of slides, *Optical Phenomena*: aurora, photo by G. V. Black; noctilucent clouds, photo by F.H. Ludlam
    (R3) MOD (RAF) photograph; British Crown Copyright Reserved
    (R4) Photo R.S. Scorer
30   U.S. Navy photograph
31   Bristol Aerojet Limited
33   Photo Emil Schulthess/ Conzett & Huber
35–6   Arthur N. Strahler, *The Earth Sciences*, Harper & Row Inc., 1963
37   (R) Swiss National Tourist Office
    (CL) Photo Chris Ridley
    (BL) Photo Paul Popper Ltd.
40   (T) After G.B. Tucker
    (B) After J.C. Johnson, *Physical Meteorology*, Technology Press of M.I.T.

and John Wiley & Sons, 1954
43   (T) Photo Eileen Ramsay
44   Photo Emil Schulthess
45   Photo The British Travel Association
47   O.G. Sutton, *Understanding Weather*, Penguin Books Ltd., 1964
48–   Arthur N. Strahler, *The*
52   *Earth Sciences*, Harper & Row Inc., 1963
53   Aero Service Corp.
54–5   (T) H.H. Lamb, *Meteorological Magazine*, December, 1960
58   Arthur N. Strahler, *The Earth Sciences*, Harper & Row Inc., 1963
60   Hydrodynamics Laboratory, Department of the Geophysical Sciences, University of Chicago
61   Photo Ivan Massar/Black Star
63   (B) Photo Georg Gerster
64   (TR & CR) Photos R.S. Scorer
    (BR) Photo R.S. Scorer
65   (T) Photos Carol Unkenholz
66   (L) Science Museum, London/photo Michael Holford
69   (T) Photo Swissair
    (BL) Photo courtesy Wright Rain, Ringwood
    (BR) Photo Maurice Nimmo
71   (T) Camera Press Ltd./photo Don Knight
72–3   (T) Photo J.H. Willink
75   Aerofilms Limited
76   (T) Photo R.S. Scorer
77   (T) Photo R.S. Scorer
    (B) From the Diana Wyllie set of slides, *Optical Phenomena*, photo by J.F. Steljes
78   Photos R.S. Scorer
79   Aerofilms Limited
81   Camera Press Ltd./photo Tom Smith
82   Photos courtesy Mrs. Ukichiro Nakaya
83   Cave Collection, Royal Meteorological Society
85   (T) Photo J. Allan Cash
86   (TL) Meteorological Office, Bracknell/photo Gayroma
    (BL) Photo John Gay
    (R) Photos courtesy Mrs. Ukichiro Nakaya
87   Photo Robert M. Cunningham
89   (T) After F.H. Ludlam
    (BL) Meteorological Office, Bracknell
    (BR) Photo F.H. Ludlam
91   (T) *Scientific American*, © 1957
    (B) After Mason
93   Photo Richard E. Orville, University of Arizona
94   Arthur N. Strahler, *The Earth Sciences*, Harper & Row Inc., 1963
95   Photo Keystone
96   From the Diana Wyllie

filmstrip *Clouds*
97 Photo Emil Schulthess
98 (L) Photo Bill Burkett,
Dallas, Texas
(R) Humble Oil & Refining
Company, Houston, Texas
99 Photo Keystone
101 Photos Michael Holford/
courtesy Director-General of
the Meteorological Office
102 (T) After S. Petterssen,
*Introduction to Meteorology*,
© McGraw-Hill Book
Company, New York, 1958
(B) A. Austin Miller and
M. Parry, *Everyday
Meteorology*, Hutchinson &
Co. (Publishers) Ltd.,
London, 1963
103 (L) Photo R.S. Scorer
(R) Aero Service Corp.
104–5 (T) After S. Petterssen,
*Introduction to Meteorology*,
© McGraw-Hill Book
Company, New York, 1958
(B) After D.E. Pedgley
107–8 After S. Petterssen,
*Introduction to Meteorology*,
© McGraw-Hill Book
Company, New York, 1958
110 U.S. Department of
Commerce, Environmental
Science Services
Administration, Weather
Bureau
111 (T) After R.S. Scorer,
*Science Journal*,
March 1966
(B) Courtesy The American
National Red Cross
112 Diana Wyllie Limited/
(L) Photo R.H. Simpson,
(R) NASA Project Mercury
113 *Life*, © Time Inc.
115 (L) Photo Ken Coton
(R) Photo R.S. Scorer
117 Engelberg Valley; photo
courtesy Swiss National
Tourist Office
119 (B) Photo Michael Holford/
courtesy East Malling
Research Centre
120–1 (T) Photo P.G. Mott
122 (T) Arthur N. Strahler, *The
Earth Sciences*, Harper &
Row Inc., 1963
(B) Photo Michael Holford/
courtesy East Malling
Research Centre
125 (T) Photo R.S. Scorer
127 (T) Aero Service Corp.
(C & B) Courtesy The
National Society for Clean
Air/photo John A. Rose
128 Photo R.S. Scorer
130 By permission of the
Controller of Her Britannic
Majesty's Stationery Office;
British Crown Copyright
131 Courtesy The Geographical
Association
133 The London Museum/photo
Michael Holford
134 *Scientific American*, © 1952
135 (T & B) Geological Survey &

Museum, London
137 (T) Photo William E. Long
(B) British Museum (Natural
History)/photo Michael
Holford
138 Lockwood Survey Corp.
140 Mansell Collection
141 From *The Eruption of
Krakatoa*, 1888; courtesy
The Royal Society
143 (T) Camera Press Ltd.
144 Camera Press Ltd.
146 (T) French National Tourist
Office
148 *Scientific American*, 1880
149 Photo by *The Times*

ARTISTS' CREDITS

Rodolph Britto: 42

Roger Durban: 43(B), 55(B), 71(B),
91(T), 94

David Litchfield: 91(B)

David Nash: 23(T), 31(T), 40(T), 48,
49, 63(T), 66(R), 72(B), 73(B), 74,
76(B), 92, 102, 104(T), 105(T), 108,
109(B), 119(T), 129, 130(B), 131, 134,
146(B)

David Parry: 12

Edward Poulton: 40(B), 41, 109(T)

E. Summers: 54–5(T), 58

Sidney Woods: 19(C & B), 25(R),
28–9, 34, 35, 36, 39, 46, 47, 50, 51,
52, 56–7, 64(L), 65(B), 77(B), 85(B),
89(T), 104(C & B), 105(C & B), 107,
111(T), 120(B), 121(B), 122(B), 125(B)